Conifers - From Seed to Sustainable Stands

Edited by Teresa Fidalgo Fonseca and Ana Cristina Gonçalves

Published in London, United Kingdom

Conifers - From Seed to Sustainable Stands
http://dx.doi.org/10.5772/intechopen.1000446
Edited by Teresa Fidalgo Fonseca and Ana Cristina Gonçalves

Contributors
Alemayehu Abera Kedanu, Amanpreet Kaur, Ana Cristina Gonçalves, Andrea Hevia, Benson Kumuli Gusamo, Carlos Pacheco Marques, Daniel Moreno-Fernández, Hana Tamrat Gebirehiwot, Iciar Alberdi, Isabel Cañellas, José Luis Louzada, Megersa Tafese Adugna, Rajesh Monga, Renato N.M. Costa, Richard Clemente, Teresa Fidalgo Fonseca

First published in London, United Kingdom, 2024 by IntechOpen
IntechOpen is the global imprint of INTECHOPEN LIMITED, registered in England and Wales, registration number: 11086078, 167-169 Great Portland Street, London, W1W 5PF, United Kingdom

British Library Cataloguing-in-Publication Data
A catalogue record for this book is available from the British Library

Additional hard and PDF copies can be obtained from orders@intechopen.com

Conifers - From Seed to Sustainable Stands
Edited by Teresa Fidalgo Fonseca and Ana Cristina Gonçalves
p. cm.
Print ISBN 978-0-85466-317-0
Online ISBN 978-0-85466-316-3
eBook (PDF) ISBN 978-0-85466-318-7

For EU product safety concerns:
IN TECH d.o.o., Prolaz Marije Krucifikse Kozulić 3, 51000 Rijeka, Croatia,
info@intechopen.com or visit our website at intechopen.com.

Meet the editors

Teresa Fidalgo Fonseca is an associate professor with habilitation at the University of Trás-os-Montes e Alto Douro (UTAD), Portugal, an integrated member of the Centre for the Research and Technology of Agro-Environmental and Biological Sciences (CITAB, UTAD), and a collaborator at the Centre of Forest Studies (CEF), University of Lisbon. She holds a Ph.D. in Forestry Sciences and conducts research in the scientific domains of her specialization, participating in national and international research projects and authoring scientific and technical publications. Her research focuses on biometrics and forest assessment, silviculture, modeling, and forest management She is Deputy of Division 1 (Silviculture) of the International Union of Forest Research Organizations (IUFRO) and coordinator of WP 1.01.10 – Ecology and Silviculture of Pine.

Ana Cristina Gonçalves is an assistant professor with habilitation in the Department of Rural Engineering, University of Évora, Portugal, and a researcher at the Mediterranean Institute for Agriculture, Environment and Development (MED). She is also the Deputy Coordinator of the International Union of Forest Research Organizations (IUFRO) subdivision 1.04.00 – Agroforestry. Dr. Gonçalves holds a Ph.D. in Forestry and has authored more than 100 publications and participated in 15 research projects. Her research focuses on silviculture and modeling in pure, mixed, even-aged, and uneven-aged stands; biomass; forest management; and planning integrated in a GIS environment.

Contents

Preface

Conifers have an important role in the world's forested landscapes, holding pivotal sway over global forest areas. Their significance extends beyond mere presence, as conifers boast a remarkable array of stand structures, silvicultural systems, and diverse yields, encompassing a spectrum of valuable products and services. Understanding these features of conifers, from the nuanced characterization of their species to the analysis of their stands, forests, and production, supports the management systems that govern their existence and contributes to preserving the age-old traditions associated with their utilization.

This book is a collection of contributions on conifer species from diverse perspectives, each delving into the multifaceted realm of conifer research. From species-specific investigations to comprehensive studies on characterization, modeling, management, genetics, ethnobotany, and the artistry of wooden construction, the chapters within this volume traverse a broad spectrum of knowledge. This volume includes six chapters:

Chapter 1: "Effect of Climate Change on Conifer Plant Species, *Juniperus procera*, and *Podocarpus falcatus*, in the Case of Ethiopia: Critical Review Using Time Series Data"

Chapter 2: "Importance of Araucariaceae for Plantation Development in Papua New Guinea"

Chapter 3: "Exploring the Biometric Traits and Potential of Radiata Pine (*Pinus radiata* D. Don) as a Non-Native Species for Sustainable Forest Systems in Portugal"

Chapter 4: "Influence of Silvicultural Operations on the Growth and Wood Density Properties of Mediterranean Pines"

Chapter 5: "Quantifying: Genetic Traits in *Pinus wallichiana* Seedlings in the Northwestern Himalayan"

Chapter 6: "Ethnobotany of Conifers in the Philippines"

The interdisciplinary nature of the contributions underscores the intricate web of relationships between conifers and their ecological contexts, emphasizing the interplay between biological, ecological, and anthropogenic factors. Editing this book has proven to be a gratifying experience, and we anticipate that it will serve as a valuable reference. We encourage researchers, academicians, and enthusiasts to delve into the evolving discourse surrounding conifers, exploring their pivotal contributions to the world's forests.

Teresa Fidalgo Fonseca
Department of Forestry Sciences and Landscape Architecture (CIFAP),
University of Trás-os-Montes and Alto Douro,
Vila Real, Portugal

Forest Research Centre (CEF),
School of Agriculture,
University of Lisbon, Lisboa,
Vila Real, Portugal

Ana Cristina Gonçalves
MED – Mediterranean Institute for Agriculture, Environment and Development,
CHANGE – Global Change and Sustainability Institute, Institute of Research and
Advanced Education (IIFA),
Department of Rural Engineering,
University of Évora,
Évora, Portugal

Section 1

Regeneration and Growth

Chapter 1

Effect of Climate Change on Conifer Plant Species, *Juniperus procera*, and *Podocarpus falcatus*, in the Case of Ethiopia: Critical Review Using Time Series Data

Hana Tamrat Gebirehiwot, Alemayehu Abera Kedanu and Megersa Tafese Adugna

Abstract

The *Juniperus procera and Podocarpus falcatus* tree species are the only indigenous conifer plants that Ethiopia has and dominantly found in dry Afromontane forests of the country. However, dry Afromontane forests are threatened by climate change. The objective of this study is to analyze the effect of climate change on the regeneration and dominance of the *J. procera and P. falcatus* tree species in Ethiopia. The regeneration status classes and importance value index score classes analysis was done along the time series. This study revealed that *J. procera* had a fair regeneration status, while *P. falcatus* exhibited an alternate regeneration status between fair and good. Not regenerating regeneration status was recorded in 2006–2010 and 2016–2020 time series for *J. procera*, while in 2011–2015 and 2021–2023 for *P. falcatus*. Regarding the importance value index score of the species, *J. procera* had the top three throughout the all-time series except in 2011–2015 which had the lowest importance value index score, whereas *P. falcatus* had the top three importance value index score status from 2016 to 2023 time series. Safeguarding these conifer species from the negative effects of climate change relies on the attention of all responsible bodies.

Keywords: *Juniperus procera*, *Podocarpus falcatus*, sustainability, regeneration status, importance value index

1. Introduction

Conifer plants are woody plants that have simple leaves, simple pollen cones, and compound or reduced ovulate cones grouped in gymnosperms. Conifer plant species are found dominantly in the major terrestrial landscapes. However, conifers have less species diversity which accounts for less than 0.3% of the species diversity from the earth's plant species [1]. Ethiopia has eight natural vegetation types based on elevation and climate gradients. From these vegetation types, the dry Afromontane and grassland complex is found in the majority of Ethiopian parts along altitudinal gradients of 1500–3000 m.a.s.l. This forest type is considered as coniferous forest [2, 3] because the warm highland part of dry Afromontane forests with 1500 to 2500 m.a.s.l of altitude range dominated by the only two co-occurring species in the country, namely *Juniperus procera* and *Podocarpus falcatus* [4, 5]. Similarly, different scholars indicated that the dry Afromontane forest of Ethiopia is a coniferous forest. For example, the dry Afromontane coniferous forest of Dodola in the Bale Mountains [6] dominantly harbor *J. procera* and *P. falcatus* [3].

On the other hand, climate change is a common environmental problem worldwide and in Ethiopia too. For example, 19 and 3% of the country's total area experienced significant decreasing and increasing trends of rainfall, respectively from 1901 to 2020 [7]. There is also a significant mean temperature increment trend over 120 years spatially and temporally ranged from 0.24 to 1.92°C and from 0.72 to 1.08°C, respectively in Ethiopia [7]. Similarly, climate change, mean maximum and minimum temperature, has increased by 0.047 and 0.028°C/year, respectively, for the period 1983–2014 in Ethiopia. However, the total rainfall has declined by 10.16 mm per annum whereas, the rainfall has declined by 2.198, 4.541, 1.814, and 1.608 mm per annum for Ethiopian summer, spring, autumn, and winter seasons respectively [8]. A slight increase in average temperature with an insignificant trend but a significant trend in minimum temperature is documented, while a decreasing trend of rainfall is documented in dry Afromontane forest fragments in northern Ethiopia [9].

Consequently, the dry Afromontane forest is highly sensitive to climate change in combination with other factors. For example, [10] revealed that a combination of climate, topographic factors, and local human disturbance controlled the stability of dry Afromontane forests. Furthermore, the dry Afromontane conifer forest, as well as the rest of the forest of the country, is at risk due to the expansion of agricultural land as a result of population pressure. For instance, [11] states that the pollen data indicated increased anthropogenic activity such as deforestation and agriculture during the last millennium in Ethiopia. Similarly, evergreen dry Afromontane forest patches in Amhara National Regional State of Ethiopia are influenced by severe anthropogenic disturbances [12]. Furthermore, [13] indicates that there is a high level of anthropogenic activities in the Bale Mountains National Park. Climate, population growth, and anthropogenic factors are the main factors that could affect montane forest ecosystems in Kenya [14]. Similarly, [15] states that climate greatly modifies the composition, structure, productivity, disturbance regimes, water production, and nutrient retention. According to combined data of plant-wax δD and δ13C values with pollen, Ethiopian highlands' vegetation is sensitive to precipitation changes [11].

However, the impact of climate change on regeneration and the dominance of coniferous species of dry Afromontane forest of Ethiopia has not been explored and reported in a detailed and holistic manner. For example, there are few studies on assessing the impact of climate change on the forest ecosystem of Ethiopia [15]. Therefore, the impact of climate change on coniferous species of dry Afromontane forest of Ethiopia namely *J. procera* and *P. facaltus* species are evaluated from the perspectives of the regeneration and dominance status along time series, and the predicted impact of climate change on their future spatial distribution. Therefore, this chapter provides a better understanding of the effect of climate change on the coniferous species of dry Afromontane forests that allows urgent and sustainable adaptation actions to enhance resilience.

DOI: http://dx.doi.org/10.5772/intechopen.1004111

2. Methodology

The data sources of this chapter were peer-reviewed published papers. The articles were searched by Google Scholar using sentences such as "impacts of climate change on the dry Afromontane forest of Ethiopia" and "climate change impact on *J. procera* and *P. falcatus* in Ethiopia." The names of each species were used separately in the searching process. Keywords such as dry Afromontane, structure, regeneration, and Ethiopia were also used in searching for the status of dry Afromontane forests in Ethiopia. Generally, 152 articles were downloaded and from these 102 were used for this work. The collected data were organized and analyzed in time series accordingly following scientific standards. Time series data are the genuine way to understand the change in ecological processes of terrestrial and aquatic ecosystems in ecology [16, 17]. Therefore, in this study time series data were used to understand the effect of climate change on the regeneration and dominancy in coniferous species of Ethiopia where the dominance of the species is analyzed from the importance value index (IVI) [18] score of the species in the forest.

Data were analyzed across time series 1996–2023 for regeneration data and 2006–2023 for IVI data. The time series was fixed based on the availability of published documents on the coniferous species of Ethiopia. The time series were classified as presented here below. Time series for regeneration data: 1996–2000, 2001–2005, 2006–2010, 2011–2015, 2016–2020, 2021–2023. Time series for IVI data: 2006–2010, 2011–2015, 2016–2020, 2021–2023. The regeneration status of a species is the potential/capacity for renewal of species in the forest community [19, 20]. The regeneration status classes were good, fair, poor, and not regenerating. The regeneration status was defined and analyzed by comparing the density of seedlings and saplings with the density of mature trees as follows [21]. Good regeneration, if the seedling is greater than the sapling and mature tree/adult (seedling > sapling > mature tree/adult). Fair regeneration, if seedling > or ≤ sapling ≤ mature tree. Poor regeneration occurs if a species survives only in the mature and sapling stages but does not have seedlings. Not regenerating, if a species is present only in an adult form. However, IVI is the sum of the species' relative density, relative frequency, and relative dominance used to describe and compare the dominance of a species in the whole plot [18]. Where relative density is the density of a particular species in relation to the total density of all species [18]. Relative frequency is the frequency of a certain species expressed as a percentage of the sum of frequency values for all species existing [18]. Relative dominance is the basal area of a given species stated as a percentage of the total basal area of all species present [18]. The species with the highest IV index score is considered the most important in a plot and this index is used to determine the general importance of each species in the community structure. The IVI score classes were the top three, the top five, the top ten, the middle, and the lowest. The regeneration status and the IVI status of the species data were analyzed using percentiles, and the results were presented using bar graphs and tables.

3. The distribution and status of conifer plant species in Ethiopia

3.1 Species descriptions

J. procera is the only juniper that grows naturally in both the northern and southern hemispheres while, all other *Juniperus* species are confined to the northern hemisphere. *J. procera* is native to the mountainous regions and highlands of Sudan, Eritrea, and Ethiopia southward through East Africa and eastern DR Congo to Malawi and Zimbabwe and also in Saudi Arabia/Yemen [22, 23]. *J. procera* found in East Africa occurs most commonly with an altitudinal range between 1800 and 2700 m, where the rainfall averages 1000–1200 mm annually. It occurs abundantly in western Kenya and in the Ethiopian highlands [24]. *J. procera*, a dioecious species with distinct male and female cones, is an afro-montane tree often reaching 30–35 m high, and can reach 50 m maximum of the largest tree of its genus. *J. procera* is a major component of the forest that is transitional between dry, single-dominant afro-montane forest and semi-evergreen bushland and thicket. *J. procera* will not regenerate in mature forests, but is replaced by *Podocarpus* forests and similar forest types (**Figure 1**) [25].

P. falcatus specie's family *Podocarpaceae* is the second largest among conifer families with incredible diversity and functional traits, and it is the dominant southern hemisphere conifer family. Furthermore, the species *P. falcatus* synonym with *Afrocarpus gracilior* is native to Ethiopia, Kenya, Tanzania, Congo, Rwanda, South Sudan, and Uganda [26]. *P. falcatus* species is naturally growing up to 45 m high and 250 cm in diameter in 11 out of the 14 floral regions recognized in Ethiopia [27]. This tree was found predominantly in undifferentiated Afromontane forests with an altitude range of 1550–2800 m, a mean annual temperature of 13–20° C, a mean annual rainfall of 1200–1800 mm, and humus-rich sandy soils [27, 28]. *P. falcatus* is a dioecious species and is a wind-pollinated species (**Figure 2**) [28].

3.2 The distribution of conifer plant species in Ethiopia

J. procera and *P. falcatus*, plant species, are found in the dry Afromontne forest of Ethiopia predominantly and rarely in the moist montane forest (**Tables 1** and 2, **Figure 3**). This is due to the warm highlands ("Woina Dega") zone of dry Afromontne forest in the altitude ranges of 1500 to 2500 m.a.s.l, temperatures of 15 to 20°C and rainfall ranges between 800 and 2400 mm is characterized by the occurrence of the only two conifers in the country. The cold and dry parts of these highlands are dominated by *J. procera*, while the moist and humid parts support *P. falcatus* [5]. Similarly, the tree density of *P. falcatus* increased with

Figure 1.
J. procera *specie. 1. Matured tree of* J. procera *from St. Gebriel Church, Fiche, Ethiopia. 2. Sapling of* J. procera *from Salale University (General Tadesse Biru Campus), Fiche, Ethiopia.*

Figure 2.
P. falcatus *specie. 3. Matured tree of* P. falcatus *from Salale University (General Tadesse Biru Campus), Fiche, Ethiopia. 4. Sapling of* P. falcatus *from Salale University (General Tadesse Biru Campus), Fiche, Ethiopia.*

No	Time series	IVI score	Status	Forest name	Vegetation type	Sources
1	2006–2010	82.04	2nd	Adelle forest	Dry Afromontane forest	[29]
2	2006–2010	23.66	top five	Boditi forest	Dry Afromontane forest	[29]
3	2006–2010	53.16	1st	Hugumbirda-Gratkhassu national forest priority area	—	[30]
4	2011–2015	32.5	1st	Menagesha Amba Mariam forest	Dry Afromontane forest	[31]
5	2011–2015	0.43	the lowest	Gedo forest	Dry Afromontane forest	[32]
6	2011–2015	1.61	the lowest	Tara Gedam forests	—	[33]
7	2011–2015	68.42	1st	Boda forest	Dry Afromontane forest	[34]
8	2011–2015	1.01	the lowest	Gendo forest	Moist evergreen montane forest	[35]
9	2016–2020	125.66	1st	Yerer mountain forest	Dry Afromontane forest	[36]
10	2016–2020	52.86	in the middle	Kumuli forest	Dry Afromontane forest	[37]
11	2016–2020	26.51	1st	Chilimo forest	Dry Afromontane forest	[38]
12	2016–2020	93.52	1st	Arero forest	Dry Afromontane forest	[39]
13	2016–2020	34.15	2nd	Ades forest (Southeastern Ethiopia)	Dry Afromontane forest	[40]
14	2016–2020	16.98	3rd	Yegof forest	Dry Afromontane forest	[41]
15	2016–2020	46.5	1st	Chilimo Gaji forest	Dry Afromontane forest	[42]
16	2016–2020	81.45	1st	Debre Libanos church forests	Dry Afromontane forest	[43]
17	2016–2020	12.2	top ten	Awi Zone of forests	Dry Afromontane forest	[44]
18	2016–2020	67.9	1st	Hugumburda forest	Dry Afromontane forest	[45]
19	2016–2020	0	the lowest	Gelawoldie community forest	Dry Afromontane forest	[48]
20	2016–2020	16.984	3rd	Yegof forest	Dry Afromontane forest	[41]
21	2016–2020	36.2	2nd	Ades forest (West Hararghe Zone1	Dry Afromontane forest	[47]

No	Time series	IVI score	Status	Forest name	Vegetation type	Sources
22	2016–2020	0.179	the lowest	Amoro forest	Dry Afromontane forest	[48]
23	2016–2020	36.9	2nd	Gatira George's forest	Dry Afromontane forest	[49]
24	2016–2020	1.992	the lowest	Gemechis forest	Dry Afromontane forest	[50]
25	2016–2020	3.643	in the middle	Weiramba forest	Dry Afromontane forest	[51]
26	2016–2020	lower	the lowest	Tore forest	Plantation forest	[52]
27	2021–2023	18.46	top five	Tulu Korma forest	Dry Afromontane forest	[53]
28	2021–2023	15.53	top ten	Harego forest	Dry Afromontane forest	[54]
29	2021–2023	154.9	1st	Hurubu forest	Dry Afromontane forest	[55]
30	2021–2023	148.5	1st	Gennemar forest	Dry Afromontane forest	[56]
31	2021–2023	(upper altitude) 43.06	1st	Werganbula forest	Dry Afromontane forest	[57]
32	2021–2023	(Edge) 32.49	2nd	Bale Mountains National Park forest	Moist evergreen montane forest	[13]
33	2021–2023	(Interior) 40.61	2nd	Bale Mountains National Park forest	Moist evergreen montane forest	[13]
34	2021–2023	6.72	in the middle	Gosh-Beret forest	Dry Afromontane forest	[58]
35	2021–2023	12.76	in the middle	Shoti forest	—	[59]
36	2021–2023	15.94	3	Menfeskidus Monastery forest	Dry Afromontane forest	[60]
37	2021–2023	41.7	2nd	Dindin forest	Dry Afromontane forest	[61]
38	2021–2023	149.5	1st	Less disturbed forest of Beyeda district	Dry Afromontane forest	[62]
39	2021–2023	136.8	1st	Moderately disturbed forest of Beyeda district	Dry Afromontane forest	[62]
40	2021–2023	149.2	1st	Highly disturbed forest of Beyeda district	Dry Afromontane forest	[62]

Table 1.
IVI status data of Juniperus procera *species.*

increasing altitude from 1500 to 1900 m.a.s.l and then decreased with the absence of mature trees at 2100 m in the Harenna forest, southeastern Ethiopia [72].

3.3 Status of conifer plant species in Ethiopia

3.3.1 Regeneration status of J. Procera *and* P. falcatus species

The regeneration status and IVI score of the species are an indicator of the species' health and sustainability, and hence of the forest ecosystem. The analysis indicated that *J. procera* had a good and fair regeneration status in equal percent in the time series of 1996–2000. However, no data was found during 2001–2005. Fair, poor, and not regenerating statuses were recorded in equal proportion in the 2006–2010 time series. Good (14.28%), fair (57.14%), and poor (28.57%) regeneration status were documented in the time series of 2011–2015. Good, poor, and not regenerating status

No	Time series	IVI score	Status	Forest name	Vegetation type	Sources
1	2006–2010	9.35	top ten	Hugumbirda-Gratkhassu National forest priority area	—	[30]
2	2011–2015	5.6	in the middle	Gendo forest	Moist evergreen montane forest	[35]
3	2011–2015	32.6	top five	Menagesha Amba Mariam forest	Dry Afromontane forest	[31]
4	2011–2015	19.62	top five	Gedo forest	Dry Afromontane forest	[32]
5	2011–2015	52.47	top three	Kimphe Lafa natural forest	Dry Afromontane forest	[63]
6	2011–2015	lower	the least	Boda forest	Dry Afromontane forest	[34]
7	2016–2020	18.21	3rd	Berbere forest	Moist evergreen montane forest	[64]
8	2016–2020	11.786	in the middle	Yegof forest	Dry Afromontane forest	[41]
9	2016–2020	24.8	3rd	Wabero forest	Moist evergreen montane forest	[65]
10	2016–2020	74.5	1st	Ades forest (West Hararghe Zone)	Dry Afromontane forest	[47]
11	2016–2020	74.15	top ten	Kumuli forest	Dry Afromontane forest	[37]
12	2016–2020	13.77	3rd	Chilimo forest	Dry Afromontane forest	[38]
13	2016–2020	49.06	1st	Ades forest (Southeastern Ethiopia)	Dry Afromontane forest	[40]
14	2016–2020	11.79	in the middle	Yegof forest	Dry Afromontane forest	[41]
15	2016–2020	42.87	2nd	Chilimo Gaji forest	Dry Afromontane forest	[42]
16	2016–2020	3.43	in the middle	Hugumburda forest	Dry Afromontane forest	[45]
17	2016–2020	1.7	the least	Coffee-based Zegie Peninsula forest	Dry Afromontane forest	[66]
18	2016–2020	0.49	the lowest	Non-coffee Zegie Peninsula forest	Dry Afromontane forest	[66]
19	2016–2020	70.29	1st	Munessa forest	Dry Afromontane forest	[67]
20	2016–2020	lower	the lowest	Tore forest	Plantation forest	[52]
21	2016–2020	50.35	1st	Asabot forest	Dry Afromontane forest	[68]
22	2016–2020	11.5	top ten	Gatira George's forest	Dry Afromontane forest	[49]
23	2016–2020	13.413	top ten	Gemechis forest	Dry Afromontane forest	[50]
24	2021–2023	17.14	top ten	Shoti forest	—	[59]
25	2021–2023	32.99	1st	Kenech forest	Moist evergreen montane forest	[69]
26	2021–2023	31.32	2nd	Tulu Korma forest	Dry Afromontane forest	[53]
27	2021–2023	48.9	top five	Hurubu forest	Dry Afromontane forest	[55]
28	2021–2023	91.5	2nd	Gennemar forest	Dry Afromontane forest	[56]
29	2021–2023	(upper altitude) 37.3	2nd	Werganbula forest	Dry Afromontane forest	[57]
30	2021–2023	(Edge) 13.44	top ten	Bale Mountains National Park forest	Moist evergreen montane forest	[13]
31	2021–2023	(Interior) 29.49	top five	Bale Mountains National Park forest	Moist evergreen montane forest	[13]

No	Time series	IVI score	Status	Forest name	Vegetation type	Sources
32	2021–2023	lower	the lowest	Tulu Lafto forest	—	[70]
33	2021–2023	lower	the lowest	Menfeskidus Monastery forest	Dry Afromontane forest	[60]
34	2021–2023	49.9	1st	Dindin forest	Dry Afromontane forest	[61]

Table 2.
IVI status data of P. falcatus *species.*

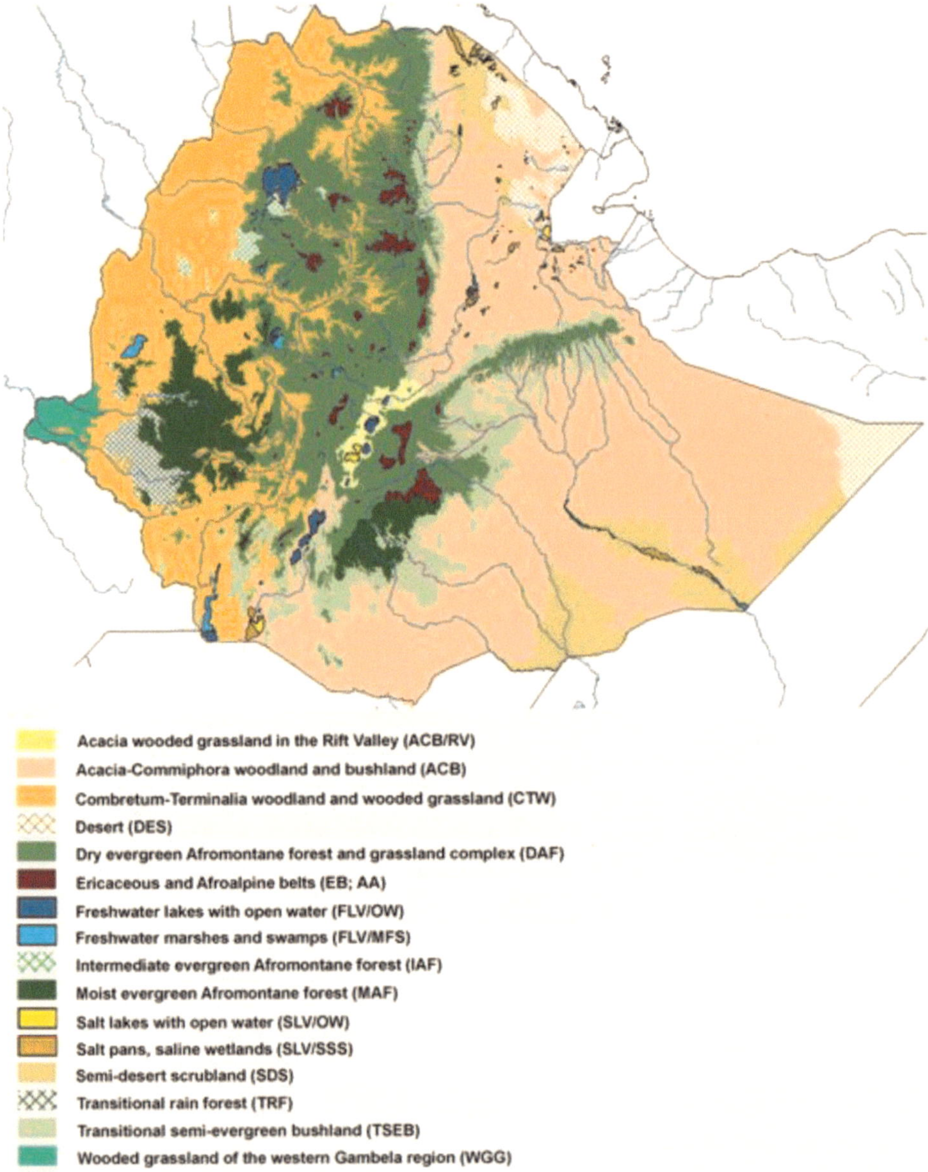

Figure 3.
Map of Ethiopian vegetation types. Source: [71].

were found in the same proportion each (20%) while, fair regeneration (40%) was found to have the highest percentage in the time of 2016–2020. *J. procera* had a good (12.5%) and fair (87.5%) regeneration status in the 2021–2023 time series (**Figure 4**). Overall, *J. procera* had the highest percentage of fair regeneration status than the other regeneration statuses from 2011 to 2015 to 2021–2023 time series.

The *J. procera* species is among the first highest density of naturally regenerated woody species with 369 individuals/ha in the case of Entoto Mountain and the surrounding area in Addis Ababa, Ethiopia, in recent times (2020) [79]. Similarly, [73] states that *J. procera* is one of the species with the highest seedling densities in Menagesha forest before 25 years. Contrary to this, [82] documented very few *J. procera* in the Wof-Washa natural forest before 28 years. Regarding soil seed bank distribution recent finding shows that *J. procera* was the third with the highest relative frequency in soil seed bank in the case of Buska Mountain in Ethiopia [83]. Recently, it has been noted that the effect of increased temperature due to climate change on the regeneration of forest species is a common problem at the global level as in the case of central Spain [84]. Nevertheless, the documented "good regeneration status" of the *J. procera* species is not satisfactory to ensure the species' healthiness and sustainability as the highest percentage is fair regeneration from 2011 to 2015 to 2023 time series. In the long run, if the regeneration status goes with a similar trend the species would be at risk.

The regeneration status analysis was also done for *P. falcatus* species. Hundred (100) percent of poor, fair, and good regeneration status were documented in the time series of 1996–2000, 2001–2005, and 2006–2010, respectively. Not regenerated (14.28%), poor (14.28%), fair (42.85%), and good (28.57%) regeneration status were documented in the time series of 2011–2015. The highest percentage in good regeneration status (77.78%) of *P. falcatus* species was observed than poor (11. 11) and fair (11. 11) in the time series of 2016–2020. Regeneration status that was not regenerated (16.67%), poor (16.67%), fair (33.33%), and good (33.33%) regeneration status were documented in the time series of 2021–2023 (**Figure 5**). Generally, *P. falcatus* species had an alternate regeneration status between fair and good from 2001 to 2005 to 2020–2023.

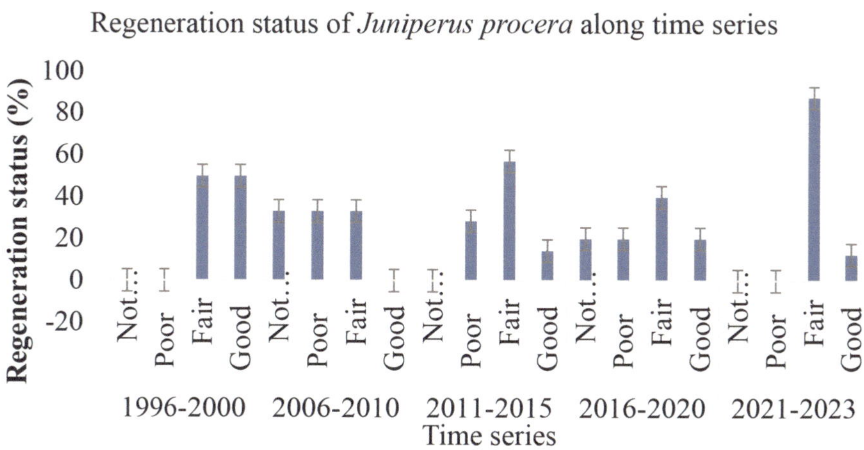

Figure 4.
Regeneration status of J. procera *along time series. Source: (see* ***Table 3****).*

J. procera species				
No	**Time series**	**Regeneration status**	**Forest name**	**Source**
1	1996–2000	Good	Menagesha forest	[73]
2	1996–2000	Fair	Gara Ades forest	[73]
3	2006–2010	Fair	Boditi forest	[29]
4	2006–2010	Not regenerating	Denkoro forest	[74]
5	2006–2010	Poor	Adelle forest	[29]
6	2011–2015	Poor	Gedo forest	[32]
7	2011–2015	Fair	Menagesha Amba Mariam forest	[31]
8	2011–2015	Fair	Chilimo forest	[75]
9	2011–2015	Fair	Borana forests	[76]
10	2011–2015	Poor	Debirelibanos Monastery forest	[77]
11	2011–2015	Good	Yegof mountain forest	[78]
12	2011–2015	Fair	Gendo moist montane forest	[35]
13	2016–2020	Good	Entoto mountain and the surrounding area forest	[79]
14	2016–2020	Not regenerating	Gedo forest	[80]
15	2016–2020	Fair	Yerer mountain forest	[36]
16	2016–2020	Fair	Kumuli forest	[37]
17	2016–2020	Fair	Chilimo forest	[38]
18	2016–2020	Poor	Arero forest	[39]
19	2016–2020	Not regenerating	Dry Afromontane forests of Awi Zone	[44]
20	2016–2020	Poor	Tore forest	[52]
21	2016–2020	Fair	Asabot forest	[68]
22	2016–2020	Good	Ades forest	[47]
23	2021–2023	Fair	Tulu Korma forest	[53]
24	2021–2023	Fair	Hurubu natural forest	[55]
25	2021–2023	Good	Werganbula forest	[57]
26	2021–2023	Fair	Dindin natural forest	[61]
27	2021–2023	Fair	Harego forest	[54]
28	2021–2023	Fair	Gosh-Beret forest	[58]
29	2021–2023	Fair	Menfeskidus Monastery forest	[60]
30	2021–2023	Fair	Gamataja Community forest	[81]

Table 3.
Regeneration status data of Juniperus procera *species.*

The *P. falcatus* species is among the top ten species with the highest seedling densities in Gara Ades and Menagesha forest before 25 years [73]. Infection of *P. falcatus* by *C. uberata* in leaves, young stems, and fruit is documented in southeastern Ethiopia and central Ethiopia that could be a threat to the regeneration of

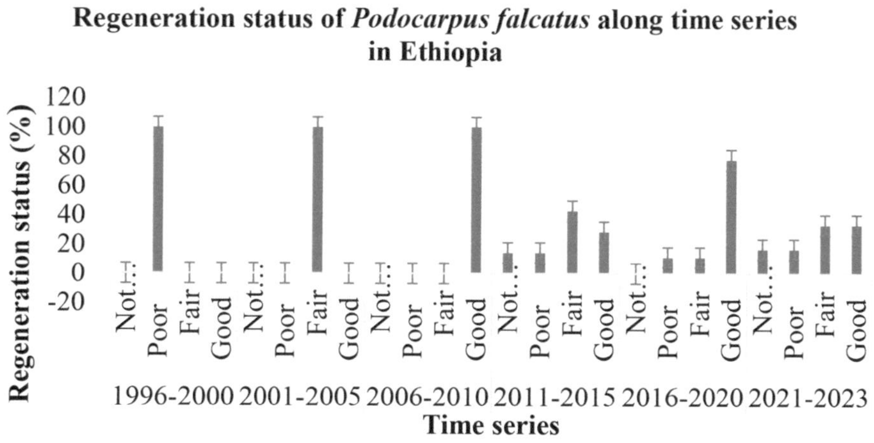

Figure 5.
Regeneration status of P. falcatus *along time series. Source: (see **Table 4**).*

***P. falcatus* species**				
No	**Time series**	**Regeneration status**	**Forest name**	**Source**
1	1996–2000	Poor	Gara Ades forest	[73]
2	1996–2000	Poor	Menagesha forest	[73]
3	2001–2005	Fair	Harena forest	[20]
4	2006–2010	Good	Munessa-Shashemene natural forest	[85]
5	2011–2015	Poor	Gedo forest	[32]
6	2011–2015	Fair	Menagesha Amba Mariam forest	[31]
7	2011–2015	Good	Debirelibanos Monastery forest	[77]
8	2011–2015	Good	Chilimo forest	[75]
9	2011–2015	Fair	Borana forests	[76]
10	2011–2015	Not regenerating	Yegof forest	[78]
11	2011–2015	Fair	Gendo moist Montane forest	[35]
12	2016–2020	Good	Kumuli forest	[37]
13	2016–2020	good	Chilimo forest	[38]
14	2016–2020	Good	Chilimo Gaji forest	[42]
15	2016–2020	Poor	Asabot forest	[68]
16	2016–2020	Fair	Munessa forest	[67]
17	2016–2020	Good	Gedo forest	[80]
18	2016–2020	Good	Berbere Afromontane moist forest	[64]
19	2016–2020	Good	Ades forest	[47]
20	2016–2020	Good	Dodola forest	[86]
21	2021–2023	Fair	Tulu Korma forest	[53]

P. falcatus species				
No	Time series	Regeneration status	Forest name	Source
22	2021–2023	Fair	Hurubu forest	[55]
23	2021–2023	Good	Werganbula forest	[57]
24	2021–2023	Good	Dindin forest	[61]
25	2021–2023	Not regenerating	Kenech forest	[69]
26	2021–2023	Poor	Gamataja community forest	[81]

Table 4.
Regeneration status data of P. falcatus *species.*

P. falcatus regeneration [87]. Furthermore, infected fruit ultimately led to the rotting of fruit and seed, which limited the seed source for *P. falcatus* regeneration of *P. falcatus* [87]. Even though the documented percent of "good regeneration status" of the *P. falcatus* species is decreasing from time to time, the documented good regeneration status does not indicate satisfactory to ensure the species' healthiness and sustainability. This is because in the time series of 2021–2023, the sum of the percentage of not regenerating and poor regeneration status is equal to good and fair regeneration status.

3.3.2 The dominance (IVI) status of the J. Procera *and* P. falcatus *species*

IVI score analysis shows that the *J. procera* scored top three, the lowest, top three, and top three classes in the time series of 2006–2010, 2011–2015, 2016–2020, and 2021–2023, respectively (**Figure 6**). This might indicate that *J. procera* tree is well adapted to the complex pressure of environmental and disturbance factors that regulate the distribution, abundance, and productivity of the species from previous to current conditions. Since [88] indicates the significant impact of altitude, aspect, slope, grazing, and human interference on species distribution and the formation of plant communities in dry Afromontane forest patches of northwestern Ethiopia. Even if *J. procera* is the dominant tree in the dry Afromontane forest of Ethiopia, it is one of the species that was observed with some stumps, few logs, and dead but standing individuals in the Denkoro forest [74].

The IVI score of *P. falcatus* was the top ten, top five, top three, and top three classes across the time series of 2006–2010, 2011–2015, 2016–2020, and 2021–2023, respectively (**Figure 7**). This indicated the increasing dominance trend of *P. falcatus* species along time series. This might be because *P. falcatus* will regenerate in matured forest and the matured forest could gradually dominated by *P. falcatus* species [25].

3.4 Effect of climate change on sustainability of conifer plant species in Ethiopia

Climate change is affecting living organism distribution in general and the effect will continue to influence the future distribution of living organisms. For example, ref. [89] indicated that all vegetation types are affected by climate changes and forests are affected by altering forest regeneration patterns, a decrease in dominance of conifer species, compositional and structural changes in forests, and upward migration of

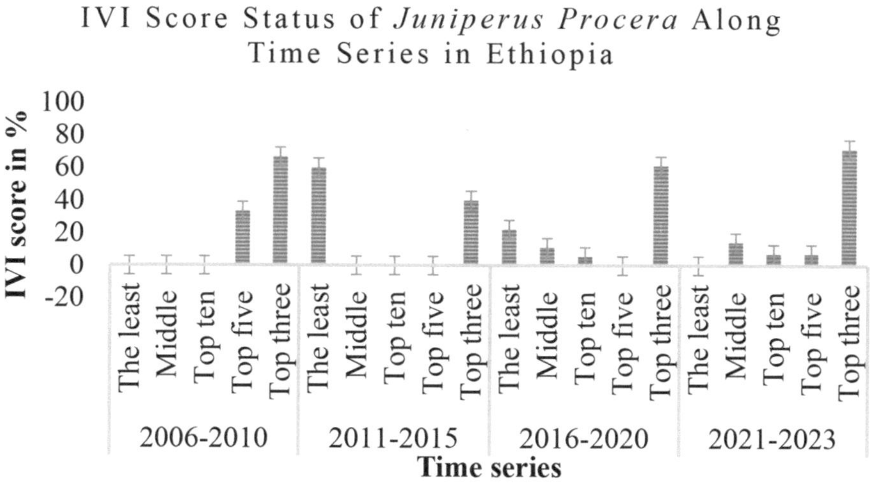

Figure 6.
IVI score status of J. procera *along time series. Source: (see* ***Table 1****).*

species in the mountains. For instance, ref. [90] states that endemic *Juniperus* species of China predicted to lose an entire of their suitable habitats due to change in temperature annual range and isothermality under full dispersal and RCP4.5 scenarios. Similar to this, suitable habitats of *J. procera* in Ethiopia will be decreased by 79.84, 91.17, 75.31, and 96.25% in Mid-century RCP2.6, Mid-century RCP8.5, End-century RCP2.6, and End-century RCP8.5 when compared with current distributions, respectively [91]. Furthermore, indicated that the annual growth of *J. procera* in Ethiopia is mainly controlled

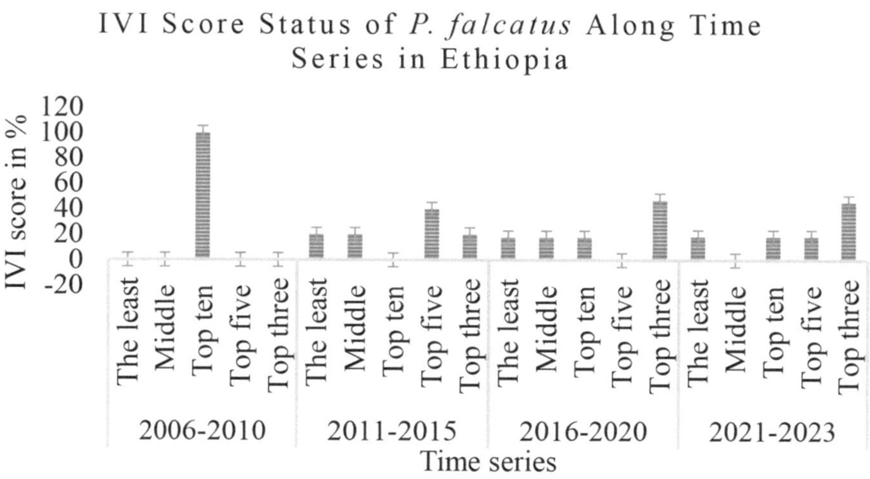

Figure 7.
Status of the IVI score of P. falcatus *along time series. Source: (see* ***Table 2****).*

by precipitation [92]. Similarly, [93] found that reduced rainfall will lead to high-level dieback of the *J. procera* species as observed in east-facing slopes than in west-facing slopes as the west-facing slope shows greener vegetation due to the aspect receiving higher rainfall in the case of Alsouda highlands, Saudi Arabia. Ref. [94] shows the poor regeneration status of *J. procera* under protected conditions after 3 years of enclosure and under open management systems in a dry Afromontane forest in northern Ethiopia, indicating that protecting the forest from livestock and human disturbance only is unlikely to lead to regeneration of this species. This might be due to moisture limitation as [95] states that poor soil moisture and nutrient conditions in dry highlands in Ethiopia result in low rates of seedling field survival and growth of native trees. Ref. [96] also states that woody plant species' seedling survival depends on both abiotic and biotic factors in an African montane forest. For instance, drought stress and potential heat stress affect the viability, growth potential, and photochemical efficiency of young *J. seravschanica* trees in the field in the case of the mountains of Oman [97].

P. falcatus was predicted to expand to higher elevations under RCP 4.5 and RCP 8.5. in the future (2070) in the case of South Africa [98]. Even though there is an environmentally suitable extensive area (>48%) in the southeastern escarpment of the main Ethiopian Rift for the *P. falcatus* species, only a small portion open-land area is practically available for rehabilitation since the area has been intensively cultivated to support the densely inhabited population [99]. From a regeneration point of view, seed germination of the *P. falcatus* species naturally occurred under the shed. For example, ref. [72] pointed out that about 74% of the seedling population of *P. falcatus* species was found in the shed and 26% in the open with a soil moisture content of between 15.6 and 27.2%, especially from 21.5 to 23.2%. Similarly, [86] recorded higher proportions of seedlings (79.45%) and saplings (72.05%) under canopy shades than in open areas with seedlings (20.6%) and saplings (27.95%). Therefore, decreased rainfall amount combined with increased temperature might influence the natural regeneration of conifer species by causing the moisture stress to the forest soil. Ref. [100] indicates positive and significant correlations when the tree-ring chronologies were compared with annual rainfall and rainfall at the main growing season but not for temperature, pointing to rainfall as the major climatic driver of plant growth in the dry Afromontane forest fragments of northern Ethiopia. Similarly, [101] shows the impact of the duration and frequency of periods of water limitation on forest structure and growth of dry tropical montane forests.

4. Conclusions and recommendations

J. procera and *P. falcatus* tree species are the only conifer plants that are found dominantly in the dry Afromontane forests of Ethiopia. However, dry Afromontane forests are sensitive to climate change mainly to decreasing rainfall and increased temperature. *J. procera* species exhibited fair regeneration status while *P. falcatus* exhibited alternating regeneration status between fair and good even in the face of climate change. IVI score of the species indicated that *J. procera* and *P. falcatus* species are dominant yet in dry Afromontane forests in the era of climate change. Overall, this result is an indicator that *J. procera* and *P. falcatus* tree species could be at risk in the long run if they continue with this trend. Therefore, thoughtful adaptation strategies should be designed and applied to dry Afromontane forests of the country to safeguard these conifer species from climate change and further degradation causes.

Specifically to *P. falcatus*, illegal felling of the preferred size of *P. falcatus* trees should be reduced and/or stopped because the presence of these big trees provides seed source and shed for the seedlings. The predicted suitable area should be set aside for the conservation of coniferous species of Ethiopia and the land use plan should be governed by suitability analysis of the area to climate change.

Furthermore, the effect of climate change on the spatial distribution of *J. procera* and *P. falcatus* should be further investigated because there are limited studies. Moreover, the effect of climate change on the soil moisture condition of dry Afromontane forests should be evaluated since the moisture condition of the soil is the critical factor that can determine the occurrence and success of natural regeneration of these species even if there are sufficient seed sources.

Conflict of interest

The authors declare that there are no known competing financial interests or personal relationships that could have appeared to influence the work reported in this chapter.

Author details

Hana Tamrat Gebirehiwot*, Alemayehu Abera Kedanu and Megersa Tafese Adugna
Forestry Department, College of Agriculture and Natural Resource, Salale University, Fiche, Ethiopia

*Address all correspondence to: hanatamrat87@gmail.com

References

[1] Gernandt D, Willyard A, Syring J, Liston A. The conifers (Pinophyta). In: Genetics, Genomics and Breeding of Conifers. Vol. 2011. St. Helier, Jersey, British Channel Islands: CRC Press; 2011. DOI: 10.1201/b11075-2

[2] Tesema AB. Forest Landscape Restoration Initiatives in Ethiopia. IUCN-EARO and WWF-EARPO. 2002

[3] Asefa M, Cao M, He Y, Mekonnen E, Song X, Yang J. Ethiopian vegetation types, climate and topography. Plant Diversity. 2020;**42**(4):302-311. DOI: 10.1016/j.pld.2020.04.004

[4] Pohjonen V, Pukkala T. *Juniperus procera* Hocht. Ex. Endl. in Ethiopian forestry. Forest Ecology and Management. 1992;**49**(1–2):75-85. DOI: 10.1016/0378-1127(92)90161-2

[5] Teketay D. Seed and regeneration ecology in dry afromontane forests of Ethiopia: II. Forest disturbances and succession. Tropical Ecology. 2005; **46**(1):45-64

[6] Hundera K, Bekele T, Kelbessa E. Floristics and phytogeographic synopsis of a dry afromontane coniferous forest in the Bale Mountains (Ethiopia): Implications to biodiversity conservation. SINET: Ethiopian Journal of Science. 2007;**30**(1):1-12. DOI: 10.4314/sinet.v30i1.18277

[7] Berihun ML, Tsunekawa A, Haregeweyn N, Tsubo M, Yasuda H. Examining the Past 120 Years ' Climate Dynamics of Ethiopia. Vienna: Springer; 2023

[8] Mekonnen Z, Kassa H, Woldeamanuel T, Asfaw Z. Analysis of observed and perceived climate change and variability in Arsi Negele District, Ethiopia. Environment, Development and Sustainability. 2017;**20**(3):1191-1212. DOI: 10.1007/s10668-017-9934-8

[9] Siyum ZG, Ayoade JO, Onilude MA, Feyissa MT. Analysis of vegetation dynamics and responses to inter-annual changes of climatic variables in dry afromontane forest fragments, Northern Ethiopia. American Journal of Geographic Information System. 2018; **2018**(5):133-144. DOI: 10.5923/j.ajgis.20180705.02

[10] Hishe H, Oosterlynck L, Giday K, De Keersmaecker W, Somers B, Muys B. A combination of climate, tree diversity and local human disturbance determine the stability of dry afromontane forests. Forest Ecosystems. 2021;**8**(1):16. DOI: 10.1186/s40663-021-00288-x

[11] Jaeschke A et al. Holocene hydroclimate variability and vegetation response in the Ethiopian highlands (Lake Dendi). Frontiers in Earth Science. 2020;**8**(December):1-14. DOI: 10.3389/feart.2020.585770

[12] Masresha G, Melkamu Y. The status of dry evergreen afromontane forest patches in Amhara National Regional State, Ethiopia. International Journal of Forestry Research. 2022;**2022**:8071761. DOI: 10.1155/2022/8071761

[13] Muhammed A, Elias E. The effects of landscape change on plant diversity and structure in the Bale Mountains National Park, southeastern Ethiopia. International Journal of Ecology. 2021;**2021**:1-13. DOI: 10.1155/2021/6628282

[14] Kibet W. Assessment of Kenya' s montane forest ecosystems: A case study on the Cherangani Hills in Western

Kenya. International Journal of Science Arts and Commerce. 2016;**1**(9):46-58

[15] Adugna Bayesa A. Impacts of climate change on the forest ecosystems in Ethiopia. American Journal of Agriculture and Forestry. 2021;**9**(6):348. DOI: 10.11648/j.ajaf.20210906.13

[16] Wauchope HS et al. Evaluating impact using time-series data. Trends in Ecology & Evolution. 2021;**36**(3): 196-205. DOI: 10.1016/j.tree.2020.11.001

[17] Ducklow HW, Doney SC, Steinberg DK. Contributions of long-term research and time-series observations to marine ecology and biogeochemistry. Annual Review of Marine Science. 2009;**1**:279-302. DOI: 10.1146/annurev.marine.010908.163801

[18] Mueller-Dombois D, Ellenberg H. Aims and Methods of Vegetation Ecology. New York: John Wiley and Sons; 1974. p. 547

[19] Duchok R, Kent K, Khumbongmayum AD, Paul A, Khan ML. Population structure and regeneration status of medicinal tree *Illicium griffithii* in relation to disturbance gradients in temperate broad-leaved forest of Arunachal Pradesh. Current Science. 2005;**89**(4):673-676

[20] Tesfaye G, Teketay D, Fetene M. Regeneration of Fourteen Tree Species in Harena Forest, Southeastern Ethiopia. The Netherlands: Elsevier; 2002

[21] Dhaulkhandi M, Dobhal A, Bhatt S, Kumar M. Community structure and regeneration potential of natural forest site in Gangotri, India. Journal of Basic & Applied Sciences. 2008;**4**(1):49-52. Available from: https://www.researchgate.net/publication/237732681%0ACommunity

[22] Adams RP. Geographic variation in the volatile leaf oils of *Juniperus procera* Hochst. Ex. Endl. Phytologia. 2013; **95**(4):269-273

[23] Bussmann RW, Paniagua-zambrana NY, Njoroge GN. Juniperus Procera Hochst. Ex Endl. C. Switzerland AG: Springer Nature; 2021. pp. 619-632

[24] Sterck FJ et al. *Juniperus procera* (Cupressaceae) in afromontane forests in Ethiopia: From tree growth and population dynamics to sustainable forest use. In: Degraded Forests in Eastern Africa. Vol. January. England & Wales, London: Routledge; 2010. pp. 291-303

[25] Negash L. A Selection of Ethiopia's Indigenous Trees: Biology, Uses and Propagation Techniques. Vol. June. Addis Ababa, Ethiopia: Addis Ababa University Press; 2010

[26] Khan R, Hill RS, Liu J. Diversity, distribution, systematics and conservation status of Podocarpaceae. Plants. 2023;**12**(1171):1-53

[27] Teketay D. Natural regeneration and Management of Podocarpus falcatus (Thunb.) Mirb. in the Afromontane forests of Ethiopia. In: Silviculture in the Tropics. London and New York: Springer Verlag Berlin Heidelberg; 2011. pp. 325-337. DOI: 10.1007/978-3-642-19986-8_21

[28] Negash L. Chapter IV Podocarpus falcatus (Thunb.) Mirb. (Podocarpaceae) (Synonym: Podocarpus gracilior Pilg.). In: A Selection of Ethiopia's Indigenous Trees: Biology, Uses and Propagation Techniques. Addis Ababa, Ethiopia: Addis Ababa University Press; 2010

[29] Yineger H, Kelbessa E, Bekele T, Lulekal E. Floristic composition and

structure of the dry afromontane forest at Bale Mountains National Park, Ethiopia. SINET: Ethiopian Journal of Science. 2008;**31**(2):103-120. DOI: 10.4314/sinet.v31i2.66551

[30] Woldemichael L, Bekele T, Nemomissa S. Vegetation composition in Hugumbirda-Gratkhassu National Forest Priority Area, South Tigray. Momona Ethiopian Journal of Science. 2010;**2**(2): 27-48. DOI: 10.4314/mejs.v2i2.57673

[31] Tilahun A. Structure and regeneration status of Menagesha Amba Mariam Forest in central highlands of Shewa, Ethiopia. Advances in Life Science and Technology. 2015;**4**(4):184. DOI: 10.11648/j.aff.20150404.16

[32] Kebede B, Soromessa T, Kelbessa E. Structure and regeneration status of Gedo dry Evergreen montane Forest, West Shewa zone of Oromia National Regional State, Central Ethiopia. Science, Technology and Arts Research Journal. 2014;**3**(2):119. DOI: 10.4314/star.v3i2.16

[33] Zegeye H, Teketay D, Kelbessa E. Diversity and regeneration status of woody species in Tara Gedam and Abebaye forests, Northwestern Ethiopia. Journal of Forest Research. 2011;**22**(3): 315-328. DOI: 10.1007/s11676-011-0176-6

[34] Fikadu E, Melesse M, Wendawek A. Floristic composition, diversity and vegetation structure of woody plant communities in Boda dry evergreen montane Forest, West Showa, Ethiopia. International Journal of Biodiversity and Conservation. 2014;**6**(5):382-391. DOI: 10.5897/ijbc2014.0703

[35] Gemechu T, Soromessa T, Kelbessa E. Structure and regeneration of Gendo moist montane forest, East Wellega Zone, Western Ethiopia. Journal of Environment and Earth Science. 2015; **5**(15):149-168. Available from: www.iiste.org

[36] Yahya N, Gebre B, Tesfaye G. Species diversity, population structure and regeneration status of woody species on Yerer Mountain Forest, central highlands of Ethiopia. Tropical Plant Research. 2019;**6**(2):206-213. DOI: 10.22271/tpr.2019.v6.i2.030

[37] Woldemariam G, Demissew S, Asfaw Z. Woody species composition, diversity and structure of Kumuli dry evergreen afromontane forest in Yem District, Southern Ethiopia. Journal of Environment and Earth Science. 2016; **6**(3):53-65. Available from: www.iiste.org

[38] Tesfaye MA, Gardi O, Blaser J. Temporal variation in species composition, diversity and regeneration status along altitudinal gradient and slope: The case of Chilimo dry afromontane forest in the central highlands of Ethiopia. World Scientific News. 2019;**138**:192-224

[39] Shiferaw W, Lemenih M, Gole TWM. Analysis of plant species diversity and forest structure in arero dry afromontane forest of Borena zone, South Ethiopia. Tropical Plant Research. 2018;**5**(2):129-140. DOI: 10.22271/tpr.2018.v5.i2.018

[40] Reshad M. Woody species richness and diversity at ades dry afromontane forest of south eastern Ethiopia. American Journal of Agriculture and Forestry. 2019;7(2):44. DOI: 10.11648/j.ajaf.20190702.12

[41] Mesfin W, Zerihun W, Lulekal E. Species diversity, population structure and regeneration status of woody plants in yegof dry afromontane forest Southeastern Ethiopia. European Journal

of Advanced Research in Biological and Life Sciences. 2018;**6**(4):20-34

[42] Mammo S, Kebin Z. Structure and natural regeneration of woody species at central highlands of Ethiopia. Journal of Ecology and The Natural Environment. 2018;**10**(7):147-158. DOI: 10.5897/jene2018.0683

[43] Koricho HH, Shumi G, Gebreyesus T, Song S, Fufa F. Woody plant species diversity and composition in and around Debre Libanos church forests of north Shoa zone of Oromiya, Ethiopia. Journal of Forest Research. 2020;**32**(5):1929-1939. DOI: 10.1007/s11676-020-01241-4

[44] Gebeyehu G, Soromessa T, Bekele T, Teketay D. Species composition, stand structure, and regeneration status of tree species in dry afromontane forests of Awi zone, Northwestern Ethiopia. Ecosystem Health and Sustainability. 2019;**5**(1):199-215. DOI: 10.1080/20964129.2019.1664938

[45] Aynekulu E et al. Plant diversity and regeneration in a disturbed isolated dry afromontane forest in northern Ethiopia. Folia Geobotanica. 2016;**51**(2):115-127. DOI: 10.1007/s12224-016-9247-y

[46] Mucheye G, Yemata G. Species composition, structure and regeneration status of woody plant species in a dry afromontane forest, Northwestern Ethiopia. Cogent Food & Agriculture. 2020;**6**(1):1823607. DOI: 10.1080/23311932.2020.1823607

[47] Atomsa D, Dibbisa D. Floristic composition and vegetation structure of Ades forest, Oromia regional state, West Hararghe zone, Ethiopia. Tropical Plant Research. 2019;**6**(1):139-147. DOI: 10.22271/tpr.2019.v6.i1.020

[48] Liyew B, Tamrat B, Sebsebe D. Woody species composition and structure of Amoro forest in West Gojjam zone, North Western Ethiopia. Journal of Ecology and The Natural Environment. 2018;**10**(4):53-64. DOI: 10.5897/jene2018.0688

[49] Ayalew A. Floristic composition and vegetation structure of Gatira George's forest in Habru Woreda in North Wollo, Ethiopia. Black Sea Journal of Agriculture. 2020;**3**(1):6-16

[50] Dawud S, Sasikumar MCJM. Floristic composition, structural analysis and regeneration status of woody species of natural forest in Gemechis District of west Hararghe zone, Oromia, Ethiopia. Journal of Biology, Agriculture and Healthcare. 2018;**8**:11-24. DOI: 10.7176/jbah/9-1-07

[51] Teshager Z, Argaw M, Eshete A. Woody species diversity, structure and regeneration status in Weiramba Forest of Amhara region, Ethiopia: Implications of managing forests for biodiversity conservation. Journal of Natural Sciences Research. 2018;**8**(5):16-31. Available from: www.iiste.org

[52] Bekele T, Abebe W. Indigenous woody species regeneration under the canopies of exotic tree plantations at Tore forest, Gelana District, Southern Oromia, Ethiopia. Biodiversity International Journal. 2018;**2**(1):1-7. DOI: 10.15406/bij.2018.02.00034

[53] Deressa D, Egigu MC, Sasikumar JM. Population structure and regeneration status of woody plant species in Tulu korma dry afromontane forest, west Shewa zone, Oromia, Ethiopia. Scientifica (Cairo). 2023;**2023**:1-9. DOI: 10.1155/2023/9964663

[54] Bogale Worku B, Birhane Hizkias E, Muhie Dawud S. Diversity, structural, and regeneration analysis of woody species in the afromontane dry forest of Harego, Northeastern Ethiopia.

International Journal of Forestry Research. 2022;**2022**:40-43. DOI: 10.1155/2022/7475999

[55] Gebirehiwot HT, Kedanu AA, Guangul AA, Adugna MT. Floristic composition, structure, and regeneration status of woody plant species in Hurubu natural forest, North Shewa, Oromia region, Ethiopia. Journal of Landscape Ecology. 2023;**16**(1):85-104. DOI: 10.2478/jlecol-2023-0005

[56] Ahmed S, Lemessa D, Seyum A. Woody species composition, plant communities, and environmental determinants in Gennemar dry afromontane forest, Southern Ethiopia. Scientifica (Cairo). 2022;**2022**:1-10. DOI: 10.1155/2022/7970435

[57] Zeleke GS, Tesfaye A, Zeleke FS. Diversity, Structure and Regeneration Status of Woody Species along Altitudinal Gradient of Werganbula Forest at Sude District, Arsi Zone. 2022. Preprint

[58] Kassa GM, Deribie AG, Walle GC. Woody species composition, structure, and regeneration status of gosh-beret dry evergreen forest patch, South Gondar zone, Northeast Ethiopia. International Journal of Forestry Research. 2023;**2023**:1-16

[59] Amenu BT, Mamo GS, Amamo BA, Doko TT. Woody species structure and regeneration status of Shoti forest, Essera district Dawro zone, SNNPRG, Ethiopia. Ukrainian Journal of Ecology. 2022;**12**(2):8-18

[60] Negesse G, Woldearegay M. Floristic diversity, structure and regeneration status of menfeskidus monastery forest in Berehet District, North Shoa, Central Ethiopia. Trees, Forest and People. 2022; 7:100191. DOI: 10.1016/j.tfp.2022.100191

[61] Lemi T, Guday S, Fantaye Y, Eshete A, Hassen N, Źróbek-Sokolnik A. Woody species composition, structure, and diversity of Dindin natural forest, south east of Ethiopia. International Journal of Forestry Research. 2023;**2023**:1-13. DOI: 10.1155/2023/5338570

[62] Taju M, Alemu A, Teshome E. Diversity, structure and regeneration status of woody species in Juniperus dominated dry afromontane forest of Beyeda district, Northern highlands of Ethiopia. IAEES Proceedings of the International Academy of Ecology and Environmental Sciences. 2021;**2021**(3):103-127. Available from: www.iaees.org Articlewww.iaees.org

[63] Aliyi NK, Hundera K, Dalle G. Floristic composition, vegetation structure and regeneration status of Kimphe Lafa natural forest, Oromia regional state, West Arsi, Ethiopia. International Journal of Biodiversity and Conservation. 2015;**5**(1):19-32. DOI: 10.5897/ijbc2018.1241

[64] Bogale T, Datiko D, Belachew S. Structure and natural regeneration status of woody plants of berbere afromontane moist forest, bale zone, south east Ethiopia; implication to biodiversity conservation. Open Journal of Forestry. 2017;7(73021):352-371. DOI: 10.4236/ojf.2017.73021

[65] Nigatu D, Firew K, Mulugeta K. Floristic composition, vegetation structure and regeneration status of Wabero forest, Oromia regional state, southeastern Ethiopia. International Journal of Biodiversity and Conservation. 2019;**11**(9):272-279. DOI: 10.5897/ijbc2018.1241

[66] Belay B, Zewdie S, Mekuria W, Abiyu A, Amare D, Woldemariam T. Woody species diversity and coffee production in remnant semi-natural dry

afromontane forest in Zegie peninsula, Ethiopia. Agroforestry Systems. 2018; **93**(5):1793-1806. DOI: 10.1007/s10457-018-0285-8

[67] Ahmedin A, Eliasb E. Tree species composition, structure and regeneration status in Munessa natural forest, southeastern Ethiopia. Eurasian Journal of Forest Science. 2020;**8**(1):21-39. DOI: 10.31195/ejejfs.622956

[68] Tura T, Soromessa T, Leta S, Argaw M. Plant community composition and structure of asabot dry afromontane forest, west Harare zone, Ethiopia. Journal of Biodiversity & Endangered Species. 2017;**05**(04):1-12. DOI: 10.4172/2332-2543.1000202

[69] Balemlay S, Siraj M. Population structure and regeneration status of woody species in Kenech forest, Southwest Ethiopia. International Journal of Forestry Research. 2021;**2021**: 1-14. DOI: 10.1155/2021/6640285

[70] Gurmessa F, Warkineh B, Soromessa T, Demissew S. Vegetation structure and regeneration status of Tulu Lafto Forest, Horo Guduru Wollega zone, West Ethiopia. SSRN Electronic Journal. 2022;**no. January**:1-10. DOI: 10.2139/ssrn.4187623

[71] Friis I, van Breugel P, Weber O, Demissew S. The Western Woodlands of Ethiopia. Copenhagen: Royal Danish Academy of Sciences and Letters; 2022

[72] Tesfaye G, Teketay D. Distribution of *Podocarpus falcatus* along environmental gradients and its regeneration status in Harenna forest, Southeastern Ethiopia author for correspondence. Ethiopian Journal of Natural Resources. 2005;**7**:118-129

[73] Teketay D. Seedling populations and regeneration of woody species in dry afromontane forests of Ethiopia. Forest Ecology and Management. 1997;**98**(2): 149-165. DOI: 10.1016/S0378-1127(97)00078-9

[74] Ayalew A, Bekele T, Demissew S. The undifferentiated afromontane forest of Denkoro in the central highland of Ethiopia: A floristic and structural analysis. SINET: Ethiopian Journal of Science. 2006;**29**(1):45-56. DOI: 10.4314/sinet.v29i1.18258

[75] Soromessa T, Kelbessa E. Interplay of regeneration, structure and uses of some woody species. Science, Technology and Arts Research Journal. 2014;**7522** (March):90-100

[76] Soromessa T. Diversity, regeneration, structure and uses of some woody species in Borana forests of southern Ethiopia: The case of Yaballo and Arero forests. Journal of Environment and Earth Science. 2015; **5**(11):2224-3216

[77] Demie G, Lemenih M, Belliethanthan S. Plant community types, vegetation structure and regeneration status of remnant dry afromontane natural forest patch within debrelibanos monastery, Ethiopia. Open Science Repository Natural Resources and Conservation. 2013;**Online, no. open-access**:e70081972. DOI: 10.7392/openaccess.70081972

[78] Mohammed S, Abraha B. Floristic composition and structure of Yegof Mountain Forest, South Wollo. Ethiopian Journal of Science and Technology. 2013;**6**(1):33-45

[79] Atinafe E, Assefa E, Belay B, Endale Y, Seta T. Floristic diversity and natural regeneration status of Entoto Mountain and the surrounding area in Addis Ababa, Ethiopia. International

Journal of Forestry Research. 2020;**2020**: 1-10. DOI: 10.1155/2020/4936193

[80] Wami FO, Tolasa T, Zuberi MI. Forest degradation: An assessment of Gedo Forest, West Shewa, Oromia Regional State, Ethiopia. Journal of Biodiversity and Environmental Sciences (JBES). 2016;**9**(October):69-78. Available from: https://www.researchgate.net/profile/MI_Zuberi/publication/307575551_Forest_degradation_An_assessment_of_Gedo_Forest_West_Shewa_Oromia_Regional_State_Ethiopia/links/57f35ac308ae91deaa590527/Forest-degradation-An-assessment-of-Gedo-Forest-West-Shewa-Oromia

[81] Abdela A, Tigist T. Woody plant regeneration status of Gamataja community forest, in Goba district, bale zone, Oromia regional state, southeast of Ethiopia. Физиология Человека. 2021; **47**(10):576-597. DOI: 10.31857/s013116462104007x

[82] Teketay D, Bekele T. Floristic composition of Wof-Washa natural forest, Central Ethiopia: Implications for the conservation of biodiversity Demel. Feddes Repertorium. 1995;**106**:127-147

[83] Bekele M, Demissew S, Bekele T, Woldeyes F. Soil seed bank distribution and restoration potential in the vegetation of Buska Mountain range, Hamar district, southwestern Ethiopia. Heliyon. 2022;**8**(11):e11244. DOI: 10.1016/j.heliyon.2022.e11244

[84] Enríquez-de-Salamanca Á. Effects of climate change on Forest regeneration in Central Spain. Atmosphere (Basel). 2022;**13**(7):1-11. DOI: 10.3390/atmos13071143

[85] Tesfaye G, Teketay D, Fetene M, Beck E. Seedling growth and survival of indigenous tree species along a light gradient in a dry afromontane forest. Forest Research and Ecology Policies. 2011;**1**:89-107

[86] Woldearegay M, Bekele T. Structure, reproductive biology, and regeneration status of *Podocarpus falcatus* (Thunb.) R. B. Ex Mirb. In Bale Mountains, Southern Ethiopia. International Journal of Forestry Research. 2020;**2020**:8825780. DOI: 10.1155/2020/8825780

[87] Assefa A, Abate D, Stenlid J. *Corynelia uberata* as a threat to regeneration of *Podocarpus falcatus* in Ethiopian forests: Spatial pattern and temporal progress of the disease and germination studies. Plant Pathology. 2015;**64**(3):617-626. DOI: 10.1111/ppa.12295

[88] Yinebeb M, Lulekal E, Bekele T. Ecological determinants in plant community structure across dry afromontane forest patches of Northwestern Ethiopia. BMC Ecology and Evolution. 2023;**23**(1):0-13. DOI: 10.1186/s12862-023-02176-0

[89] Hufnagel L, Garamvölgyi Á. Impacts of climate change on vegetation distribution No. 1: Climate change induced vegetation shifts in the palearctic region. Applied Ecology and Environmental Research. 2014;**12**(2): 355-422. DOI: 10.15666/aeer/1101_079122

[90] Dakhil MA, Halmy MWA, Hassan WA, El-keblawy A. Endemic *Juniperus montane* species facing extinction risk under climate change in Southwest China: Integrative approach for conservation assessment and prioritization. Biology (Basel). 2021; **10**(1):63

[91] Abrha H, Birhane E, Hagos H, Manaye A. Predicting suitable habitats of endangered *Juniperus procera* tree under climate change in Northern Ethiopia. Journal of Sustainable Forestry. 2018;

37(8):842-853. DOI: 10.1080/10549811.2018.1494000

[92] Sass-Klaassen U, Couralet C, Sahle Y, Sterck FJ. Juniper from Ethiopia contains a large-scale precipitation signal. International Journal of Plant Sciences. 2008;**169**(8):1057-1065. DOI: 10.1086/590473

[93] Warrag EI, Mallick J, Singh RK, Khan RA. “Status of dieback of dieback of *Juniperus procera* (African pencil cedar) in natural stands and plantation in Alsouda highlands, Saudi Arabia”. Applied Ecology and Environmental Rsearch. 2019;**17**(2):2325–2338

[94] Aynekulu E, Denich M, Tsegaye D. Regeneration response of *Juniperus procera* and olea europaea subsp cuspidata to exclosure in a dry afromontane forest in Northern Ethiopia. Mountain Research and Development. 2009;**29**(2):143-152. DOI: 10.1659/mrd.1076

[95] Asmelash F, Rannestad MM. Challenges and strategy for successful restoration of dry evergreen afromontane forests of Ethiopia. Физиология Человека. 2021;**47**(4):124-134. DOI: 10.31857/s013116462104007x

[96] Abiem I, Kenfack D, Chapman HM. Assessing the impact of abiotic and biotic factors on seedling survival in an African montane forest. Frontiers in Forests and Global Change. 2023;**6**(February):1-11. DOI: 10.3389/ffgc.2023.1108257

[97] Al Farsi KAAY, Lupton D, Hitchmough JD, Cameron RWF. How fast can conifers climb mountains? Investigating the effects of a changing climate on the viability of *Juniperus seravschanica* within the mountains of Oman, and developing a conservation strategy for this tree species. Journal of Arid Environments. 2017;**147**:40-53. DOI: 10.1016/j.jaridenv.2017.07.020

[98] Twala TC, Fisher JT, Glennon KL. Projecting podocarpaceae response to climate change: We are not out of the woods yet. AoB Plants. 2023;**15**:1-14. DOI: 10.1093/aobpla/plad034

[99] Tesfamariam BG, Gessesse B, Melgani F. MaxEnt-based modeling of suitable habitat for rehabilitation of Podocarpus forest at landscape-scale. Environmental Systems Research. 2022;**11**(1):4. DOI: 10.1186/s40068-022-00248-6

[100] Siyum ZG, Ayoade JO, Onilude MA, Feyissa MT. Climate forcing of tree growth in dry afromontane forest fragments of northern Ethiopia: Evidence from multi-species responses. Forest Ecosystems. 2019;**1**(7):1-17. DOI: 10.1007/s42452-019-0803-y

[101] Hiltner U, Bräuning A, Gebrekirstos A, Huth A, Fischer R. Impacts of precipitation variability on the dynamics of a dry tropical montane forest. Ecological Modelling. 2016;**320**:92-101. DOI: 10.1016/j.ecolmodel.2015.09.021

Chapter 2

Importance of Araucariaceae for Plantation Development in Papua New Guinea

Benson Kumuli Gusamo

Abstract

This article reviews the potential of *Araucaria hunsteinii, Araucaria cunninghamii*, and *Agathis robusta* for developing plantations in Papua New Guinea (PNG). The species are propagated from recalcitrant (*A. hunsteinii*) and orthodox (*A. cunninghamii* and *A. robusta*) seeds. The viable seeds are extracted from ripened cones and kept in controlled rooms to maintain seed quality. The seeds are raised in nurseries for seedlings and transplanted in the field. The Bulolo and Wau plantations (PNG) are managed on a 30–40-year cutting cycle. Silviculturally, a 3 × 4 m spacing (12 m^2/tree) is applied with 833 trees/ha as initial stocking. Tending is executed in the initial stages and two medium- to high-intensity thinning operations are employed to boost the growth of residual stands. Also, synthetic fertilizers and termiticides are applied to enhance plant growth and control termite infestation in young plantations, respectively. Next, non-commercial (low) thinning is scheduled at 5–7 years with 416 stems/ha stocking (estimated volume 5.522 m^3/ha) followed by commercial (crown) thinning conducted at 17–20 years with 208 stems/ha stocking (estimated volume 17.790 m^3/ha). The 208 stems/ha is maintained as final crops up to 30–40 years with expected 30.206 m^3 volume.

Keywords: evergreen coniferous species, *Araucaria hunsteinii*, *Araucaria cunninghamii*, *Agathis robusta*, recalcitrant seeds, orthodox seeds, plantation silviculture, reforestation/afforestation

1. Introduction

The Araucariaceae is a member of evergreen coniferous trees (softwoods). There are several species of importance belonging to this family, which are confined to Asia and the Southern Hemisphere [1, 2]. Some members of Araucariaceae and their natural distributions in Asia and the Southern Hemisphere are given (**Table 1**). In New Guinea (Papua New Guinea (PNG) and West Papua of Indonesia), the three species (*Araucaria hunsteinii* K. Schum., *A. cunninghamii* Ait. ex D.Don and *Agathis robusta* F.M. Bailey var. nesophyla Whitmore) of Araucariaceae are found in the natural forests [1, 2, 15–17]. According to Havel [2], these conifers are mostly found in the lower and mid-montane rain forests between 660 and 2300 m above sea level. The species occur on the mountains of mainland New Guinea commonly in softer

Scientific name	Common name	Natural distribution	Source(s)
Araucaria hunsteinii K.Schum.	Klinkii pine	New Guinea*	[2]
Araucaria cunninghamii Ait.	Hoop pine	New Guinea*	[2–4]
Araucaria angustifolia (Bertol.) Kuntze.	Parana pine	Brazil, Chile	[5, 6]
Araucaria bidwillii Hook	Bunya pine	North Queensland, Australia	[3, 7, 8]
Araucaria heterophylla (Salisb.) Franco	Unknown	Norfolk Island	[3]
Araucaria araucana (Molina) K.Koch	Araucaria	Argentina, Chile	[9]
Agathis robusta (F.Muell.) F.M. Bailey	Kauri pine	New Guinea*, Fraser Island, Australia	[1, 4, 10]
Agathis microstachya J.F.Bailey & C.T.White	Bull kauri	Australia	[10]
Agathis atropurpurea B. Hyland	Blue kauri	North Queensland, Australia	[7]
Agathis alba (Lam.) Foxw.	Dammar pine/Kauri pine	Moluccas, Philippines, Sulawesi	[2, 6]
Agathis macrophylla (Lindl.) Mast.	Pacific kauri	Solomon Island, Vanuatu, Fiji	[3, 11]
Agathis borneensis Warb.	Borneo kauri	Indo-China, Sarawak, Malaysia, Borneo	[12]
Agathis australis (D.Don) Lindl.	New Zealand kauri	New Zealand	[1, 13]
Agathis montana de Laub.	Mt. Panie kauri	New Caledonia	[5, 14]

**PNG and West Papua of Indonesia.*

The scientific names, common names, and synonyms are checked and confirmed with International Plant Names Index [6].

Table 1.
Some members of Araucariaceae and their natural distributions in Asia and The Southern Hemisphere.

slopes and ridges in their natural habitat. These conifers have excellent wood properties and are among the top-listed timbers with other tropical broad-leaved species (hardwoods). Due to wood properties and economical value of the *Araucaria* spp. and *Agathis robusta*, the species are highly sought by the timber industries. Thus, the government via the PNG Forest Authority (PNGFA) restricted the export of round logs of these timbers since 1980 unless logs are converted into semi-finished or finished wood products [18].

The *Araucaria* spp. and *Agathis* spp. are propagated solely from seeds. The seeds from ripened cones are dispersed by wind and animals in the natural habitat [3]. The only threat to seeds on the forest floor or during storage are rodents who feed on them for food [17, 19]. The seeds of these coniferous species (*Araucaria* spp. and *Agathis robusta*) have low dormancy and can lose their viability quickly in natural conditions. For example, seeds of *A. hunsteinii* are dead within 8 weeks if left in the open and allowed to dry [19]. Also, seedling survivals are low in prolonged shadings under the forest canopy [3]. However, if there is no damage by rodents or infestations from insect pests and diseases on the fertile seeds, the species have good natural regeneration capabilities when exposed to environmental conditions (moisture, light, and temperature) conducive for germination. The species can easily be domesticated from their wildings either from seeds or from seedlings plucked up from the forest floor. Thus, these species are widely domesticated and cultivated on monoculture or mixed plantations and as ornamentals in the tropical environments of PNG and northern Australia. For silvicultural trials and plantation establishments, the seeds are selectively sourced from healthy parent trees with good vigor and stem forms. The seeds are stored in a controlled storage shed and raised in nurseries for seedling production until they reach a transplantable size for outplanting in the field. As far as timber processing and utilization are concerned, the softwoods are non-porous with desirable wood characteristics, that is, physical, mechanical, and working properties in terms of strength-to-weight ratio, sawing, peeling, gluing, seasoning, machining, and permeability to treatment with chemical solutions [16, 20].

The PNG government has developed a policy on reforestation for sustainable forest management due to depletion of forest resources as a result of industrial logging, forest clearance (deforestation) for agri industries, and shifting cultivations [21, 22]. The envisaged target is to achieve 800,000 ha of planted forest by 2050 via reforestation and afforestation activities of logged-over concessions and anthropogenic grasslands or degraded lands, respectively [23]. In the reforestation and afforestation programs, the softwoods *Araucaria* spp. and *Agathis* spp. (as well as other indigenous hardwoods of commercial value) are identified as prime candidate species for increasing the plantation capacity in the country. Apart from the production of a forest through plantation establishment, the reforestation and afforestation practices should address other sustainable development goals such as carbon sequestration and climate change mitigation.

This work provides a review on the three native coniferous species (*Araucaria hunsteinii, A. cunninghamii* and *Agathis robusta*) of Araucariaceae of PNG. Specifically, the article highlights the species' natural distributions, flowering and seed characteristics, seed storage and viability, nursery techniques for seedling production, basic silvicultural practices applied in Bulolo and Wau plantations (PNG), and the importance (potential) of the species for developing plantations.

2. Botanical description and natural distribution of *Araucaria* spp. and *Agathis robusta* in Papua New Guinea

The *Araucaria* spp. and *Agathis* spp. come from the Araucariaceae family in the order Araucariales. The Araucariaceae is classified as seed-bearing (vascular) conifers belonging to Phylum Gymnospermae [24]. Unlike the angiosperms, the gymnosperms bear arborescent seeds with no enclosing carpellary structure [25]. The *Araucaria* spp. and *Agathis* spp. have male and female flowers (unisexual) in the crown and bear naked seeds in cone-shaped fruits [2]. Whitmore [4] said that the species have characteristics of growing huge in size (50–90 m tall and 2.0–2.5 m diameter) with straight and cylindrical bole as emergent trees in the natural forest canopy (**Figure 1**). The descriptions of tree taxonomy (based on natural stands) and natural distributions of *Araucaria* spp. and *Agathis robusta* in PNG (**Table 2**) are summarized from various literature sources [2, 15–17].

The genus *Araucaria* composes of 19 species and occurs along the eastern coast of Queensland (Australia), eastern half of New Guinea, New Caledonia, Norfolk Island, southern and eastern Chile, Argentina, and southern Brazil [3, 26]. On the other hand, the *Agathis* genus comprise about 20 huge timber trees [27] that have natural distributions in the Philippines, Malaysia, New Guinea, Australia, and New Zealand

Figure 1.
Emergent trees of Araucaria hunsteinii with monopodial crown in a mixed forest of Bulolo.

Taxonomical descriptions	***Araucaria hunstenii* (Klinki pine)**	***Araucaria cunninghamii* (Hoop pine)**	***Agathis robusta* (Kauri pine)**
Habit/ Characteristics	A very tall tree (93 m height & 2 m diameter) with straight cylindrical bole, which continues to the tip. Branches whorled, horizontal, slender, with long leaf-bearing twigs crowded at the end. Young trees have a pyramidal crown but may change to round or flat tops in older trees.	Tall tree (66 m tall & 2 m diameter) with a straight cylindrical bole, which continues to the tip. Branches whorled, slender, horizontal, with leaf-bearing twigs along the whole length. Crown mostly irregular & sharply pointed in young trees.	A large tree (50 m tall & 2.5 m diameter) with a huge cylindrical bole & broad crown. Young trees with narrow pointed crowns.
Bark	Thickness: 3 cm. Outer bark: dark brown with large pustules & fissures, peeling off in thick corky flakes. Inner bark: red to pink, fibrous near the wood tissue, with thick, white resinous exudate.	Thickness: 3 cm. Outer bark: dark red-brown, rough, peeling off in papery layers, which often cling to the bole. Inner bark: brown & white, mottled, with thick, white resinous exudate.	Thickness: 2 cm. Outer bark: rusty brown, craterous & pustular, peeling off in rounded flakes. Inner bark: pink with thick, white exudate, corky, non-fibrous.
Wood	Straw colored with pink tinge, soft & light, non-porous with fine rays.	Straw colored, fairly soft & light, non-porous with fine rays & uniform texture.	Pink to light brown, soft & light, non-porous with narrow rays & fine texture.

Taxonomical descriptions	***Araucaria hunstenii* (Klinki pine)**	***Araucaria cunninghamii* (Hoop pine)**	***Agathis robusta* (Kauri pine)**
Leaves	On shaded young trees: distichous, 3 cm long, relatively thin & lanceolate. On matured trees: crowded around the branches, thick, coriaceous, broadly lanceolate & 10 cm long. Both surfaces are glossy, medium to dark green.	On shaded young trees: spreading almost distichous, 2 cm long, sharp & pointed. On matured trees: shorter (1 cm) crowded, overlapping & awl shaped.	Sub-opposite. Almost without petioles, narrowly elliptical (8 × 2 cm) thick coriaceous, parallel venation, glabrous light green. The terminal bud is rounded & has pointed buds.
Flowers	Male flowers on lower branches, in pendulous spikes (15 cm long), consists of papery scales covering anthers; light green in color. Female flowers on upper branches, in short spikes & consists of pointed scales covering the ovules.	Male flowers on thin hanging branches in the lower part of the crown, in spikes (5 cm long). Cylindrical, composed of scales & anthers. Female flowers erect on the upper-most branches in small (2 cm long) spikes, consists of sharp pointed scales & ovules.	Male flowers in axillary spikes. Spikes ovoidal to cylindrical, consists of papery scales & anthers. Female flowers in short spikes, consists of blunt scales & ovules.
Fruit	A large cone, broadly ovoid (20 cm long & 12 cm broad), consists of numerous leathery, winged scales, with sharp points. Seeds (2 cm long) are contained in scale, with broad tip & pointed base, starchy.	An ovoid cone (7 cm diameter) composed of woody, winged scales, with sharp points, which contain the seeds.	A near-globose cone, which consists of spirally arranged woody scales & seeds. Scales blunt, seeds separate from scales, thin, with coat extended into 1 or 2 wings.
Habitat	In the mountains of Eastern New Guinea, especially on softer slopes of lower montane rain forests. Mostly occurs between 660 and 1650 m above sea level.	Throughout the mainland New Guinea & islands of Milne Bay Province, mainly on the ridges of upper (mid) montane rain forests. Mostly occurs between 1000 and 2300 m above sea level.	In the mountains of mainland New Guinea & New Britain island (Pomio). Occurs mostly between 660 and 1980 m above sea level.
Natural distribution (PNG)	It is found in Bulolo, Watut, and Waria Districts (Morobe Province); border valleys of Whagi (Western Highlands) & Jimi (Jiwaka); and borders of Eastern Highlands & Morobe Provinces.	Heads of Sepik River (Ambunti, East Sepik Province).	Inland Pomio (East New Britain Province). Occurs at 2000 m altitude in Watut, Morobe Province.

Table 2.
Taxonomical/botanical descriptions and natural distributions in PNG.

[3, 17, 26]. According to Whitmore [27], the genus *Araucaria* (18 species) and *Agathis* (13 species) compose the Araucariaceae, whereby the *Agathis* spp. has wide distributions than the *Araucaria* spp. In the tropical rain forest of the Far East (**Table 1**).

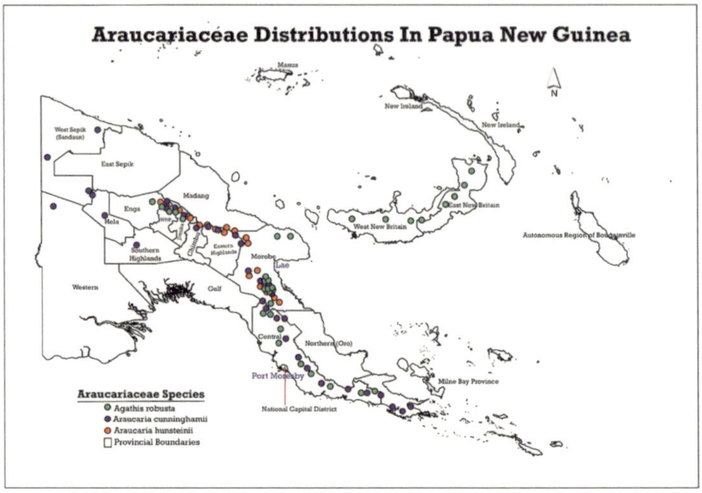

Figure 2.
Map showing the distribution of three species of Araucariaceae in Papua New Guinea.

Figure 3.
A stand of Araucaria cunninghamii at Bulolo University College.

The three coniferous species *Araucaria hunsteinii*, *A. cunninghamii*, and *Agathis robusta* are endemic in PNG, and their distributions are indicated on the map (**Figure 2**). *A. hunsteinii* grows at altitudes between 500 and 2100 m in primary forests and are distributed mainly in Bulolo, Wau, Watut, and Waria (Morobe Province); Jimi Valleys (Jiwaka Province); and Central and Western Highlands Provinces (**Figure 1**). Meanwhile, *A. cunninghamii* is scattered along the eastern coast of Australia (northern New South Wales to north-eastern Queensland) and New Guinea (**Figure 3**). In PNG, *A. cunninghamii* occurs at altitudes between 600 and 1500 m often in sub-montane forests on leached soils in association with *Podocarpus neriifolius, Prumnopitys amara,* and *Castanopsis acuminatissima.* On the other hand, *Agathis robusta* occurs at altitudes between 700 and 2000 m in mainland New Guinea, for example, Watut in Morobe Province, and in New Britain Island, for example, Pomio in East New Britain Province [2, 15, 17]. The cultivated stands of *A. robusta* are shown (**Figures 4** and 5).

Figure 4.
A stand of Agathis robusta at Bulolo University College.

Figure 5.
An unthinned plot of Agathis robusta at Bulolo plantation.

3. Flowering and fruiting

The flowering and fruiting seasons and the characteristics of the three pines (*Araucaria hunsteinii*, *A. cunninghamii* and *Agathis robusta*) differ between the species and between different geographical localities. The species are reproduced sexually via pollination mainly by the wind. For instance, in *Araucaria hunsteinii*, the pollen from the male flowers on lower branches in pendulous spikes are carried by the wind to the ovules of the female flowers on the upper branches on short spikes [16]. In Bulolo and Wau, flowering begins from January to March, and the cones mature at the end of dry season (September to October) per year. On average, a single cone weighs *ca.* 850 g with 117 viable seeds, which produces approximately 5000–6000 seeds per kilogram (**Figure 6B**). Meanwhile in *A. cunninghamii*, flowering occurs between March and June every year. The fruit, in a single cone, matures from October to December for

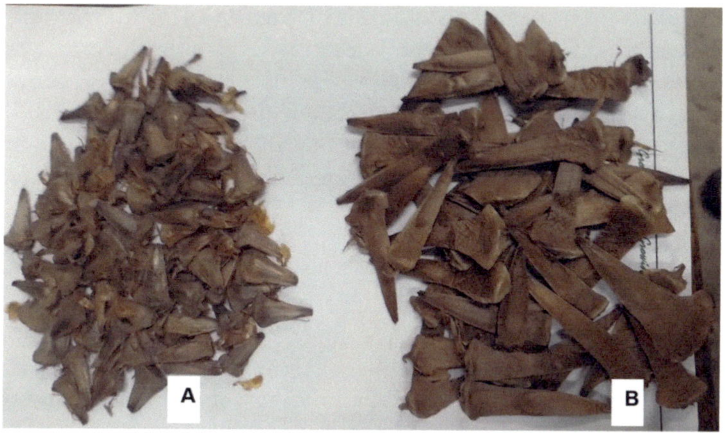

Figure 6.
Clean viable seeds: (A) Araucaria cunninghamii and (B) Araucaria hunsteinii.

collection. The mature cone weighs *ca.* 200 g with 260 seeds, which yields between 4000 and 5000 seeds per kilogram [17]. In addition, Laufenfels [3] noted that individual trees of *Araucaria* spp. can produce >500 cones per year. Later in the year, the cones become ready (ripen), and winged seeds are dispersed early in wet seasons (**Figure 6A**). In contrast to 260 seeds produced per mature cone [17], Laufenfels [3] reported 800 seeds contained in a cone. The difference in the number of seeds produced per cone is suggested to be due to the size of cones from which seeds were extracted during the time of seed collection. In this case, the 800 seeds may have been extracted from larger cones sourced from the natural forest (old-growth) stands in the 1950s [3] compared to 260 seeds derived from the cones of plantation stands [17].

Figure 7.
Clean viable seeds of Agathis robusta.

There is insufficient information available on the flowering and fruiting seasons of *Agathis robusta* in PNG. Whitmore [27] described *Agathis* spp. As monoecious, but female cones are produced before the males. The male strobili are more or less sessile and comprised of a number of bracts that produce the microsporophylls. The female cones are heavily globose and borne on shoot tips with blunt scales, each having a single ovule. At maturity, the cones are shattered where flattened ovoid seeds with wings are dispersed (**Figure** 7).

4. Seed collection, storage, and viability

In PNG, seeds of Araucariaceae are collected from seed trees of natural stands marked out in the field. Often these trees lose their vigor and health, and the ability for seed production declines over time. In order to maintain improved and quality seed sources, the best individual stands (healthy and defect-free) of *Araucaria* spp. and *Agathis robusta* are marked and selected as seed trees in few compartments of the plantations. Usually, thinning is conducted to remove the inferior layer of the stand and allow the superior layer of the stand to grow as seed trees for improving seed production and sourcing quality seeds. These seed trees in the compartment are not managed as part of a seed orchard as the seed trees are harvested during clear-cut operations. And, new seed trees are marked in established compartments. This is unlike other countries where seed orchards are permanently established and managed from genetically improved stands for the mass production of quality seeds [28].

4.1 Preparation of *Araucaria hunsteinii* seeds

The time for cone maturity and seed shedding is usually short; thus, it is important to understand cone maturity and plan seed collection promptly on time. The following methods are used to determine cone maturity: (1) the embryo should be 16 cm long with a well-developed and hard endosperm; (2) cut the tip of the cone to observe the brown color on the cone's scale; and (3) use weight, as mature cones are usually lighter than immature cones. Cones are collected immediately using hooks attached to a bamboo pole, or use ladders to climb and collect cones from orchards or parent trees. Cones are transported in copra sack bags to an extraction shed and dried for 2–3 weeks on open racks or stored in bags for a week (note: longer storage in bags can result in fungal infestations). As the cone dries, it disintegrates into individual scales, making it easy for hand separation. The seed is de-winged by hand and dried for a week before storage. The matured seeds are recalcitrant with >53% moisture content (MC) and can tolerate desiccation below 32% MC during storage (**Figure 6B**). Seed viability reduces to zero after 4 weeks when stored at 25°C; however, viability can be maintained for 6–18 months if seeds are kept in air-tight containers at a constant temperature of 3–6°C [17]. For example, Evans [19] reported seed viability up to 8 weeks, while Havel [29] observed 50% seed survival for 18 months when stored in moist conditions in a sealed container. Next, McKinty [11] added that the seeds of *A. hunsteinii* lose viability at a rate of 10–12% per week.

4.2 Preparation of *Araucaria cunninghamii* seeds

The matured cones are collected before they open while on the trees. A hook is attached to a pole to dislodge the cones. Cone maturity is determined by cutting slightly the tip of the sample cone with a sharp knife where a dark gray–brown color indicates maturity. Also, full-sized dark green cones are considered mature and are collected. The cones are then air-dried by spreading on trays for 2–3 weeks under a heavy shade. The wings are moved with hands using gloves and a wire screen. Other cleaning techniques are winnowing or the use of domestic electric fans to separate light materials from the seeds. The seeds are orthodox unlike their counterpart *Araucaria hunsteinii* (**Figure 6A**). The matured seeds have 23% MC at the initial stage and can dry to 2% MC without damage. Seeds are stored best in a refrigerator (−18°C) where they can be stored for up to 6 years. Meanwhile, seeds placed in copra sack bags or in air-tight containers can be store at −10°C for up to 3 years. Additionally, the fresh seeds have 75–80% viability [11, 17, 29].

4.3 Preparation *Agathis robusta* seeds

There is little information available on its flowering/fruiting phenology, seed preparation, and viability in PNG. According to PNGFA's National Tree Seed Centre (Bulolo, Morobe Province), *A. robusta* flowers every 6 months per year, and fruits are harvested from January to March in the same year. Tree climbers use ladders to climb the parent tree and use hooks to dislodge the cones. The collected fruits are dried in the open where cone scales are disintegrated, and seeds fall off (**Figure 7**). The seeds are separated by hand and winnowed before storage. The orthodox seeds are best stored in a refrigerator at a temperature of 3.5°C immediately as the seeds lose viability rapidly at ambient temperatures. Elsewhere in New Zealand, there could be documentation on the *Agathis australis* with regard to seed collection, preparation, and storage techniques as well as tests for viability [3, 5, 12, 27].

5. Plantation development and management with reference to Bulolo and Wau projects

Iverson et al. [30] reported that the plantation of *Araucaria* spp. began in Bulolo and Wau districts of Morobe Province after the gold discovery and extraction in the 1960s in Bulolo Valley during the Australian colonial administration of the Territory of New Guinea. As the gold deposit was depleting, a Canadian gold mining company (Bulolo Gold Dredging Ltd), which was renamed Commonwealth New Guinea Timbers Ltd. (CNGT), ventured into the timber-processing industry. The pure stands of *Araucaria* spp. in the Bulolo Valley were harvested for veneer production for plywood manufacture in the 1950s. The plantations of *Araucaria* spp. in Bulolo and Wau districts began in the 1950s, and the plantation capacity increased in the 1960s in order to maintain the raw material supply and sustain the plywood-making industry. CNGT was later changed to PNG Forest Products Ltd. in the 1970s and established as a sawmill complex (alongside the existing plymill) where plantation raw materials (*Araucaria* spp.) were fed for sawn boards and pole productions while maintaining plywood making [30].

Presently, the state through the PNGFA owns and manages 12,000 ha of pine plantations at Bulolo and Wau districts of Morobe Province. The plantations are composed of three major species: *Araucaria cunninghamii*, *A. hunsteinii* and *Agathis robusta* (including exotic *Pinus carrbaea*). A few area (ha) is planted with *Agathis robusta* within the plantations due to inadequate knowledge about its silviculture and longer cutting cycle as compared to *Araucaria* spp. However, *A. robusta* has huge potential for commercial cultivation on a plantation scale for veneer and lumber production given the favorable environmental conditions in PNG. Apart from Bulolo and Wau plantations, a few compartments of *Araucaria cunninghamii* can be found in Lapegu and Kainantu state-owned plantations in the Eastern Highlands Province (PNG). According to the PNGFA's Corporate Plan 2021–2030, the annual target of the plantation management is to raise 500,000 seedlings of *Araucaria* spp., reforest 330 ha, and harvest 100,000 m^3 volume of raw materials [23]. The PNGFA regards Bulolo and Wau monoculture plantations as a model project, and the project anticipates to achieve sustainable yield and high economic return on investment with prudent management and application of appropriate silvicultural practices.

5.1 Nursery techniques: propagation and management

The Bulolo and Wau plantations' nursery is designed to hold 800,000 plus seedlings of high quality (premium 'A' grade) at any one time per annum (**Figures 8** and **9**). Some specific tasks of the nursery management are to: (1) ensure seed tree locations are well-defined and take necessary measures to protect them (in order to protect gene erosion); (2) supervise collection, extraction, drying, and storage of seeds; (3) plan and implement nursery production of healthy seedlings to meet the target and on schedule; and (4) ensure high quality (disease-free) seedlings are dispatched for reforestation and afforestation activities. The major nursery activities undertaken in order to produce high quality materials for plantation establishments include seed and soil collections, seedling production, and seedling maintenance [17, 28].

Seed collection. Seeds are collected from parent trees marked out in the Bulolo and Wau plantations. The viable seeds are stored at PNGFA's National Seed Centre, Bulolo.

Soil collection. The soils for nursery are usually top layer (humus) collected from the mid-montane forest. The soils are sieved and sterilized using fire to eliminate

Figure 8.
Araucaria cunninghamii seedlings stored under a shed in stand-out beds, Bulolo plantation.

Figure 9.
Saloon cloth is removed for seedlings to undergo hardening or culling. Seedlings are ready for out-planting in the field, Bulolo plantation.

unwanted microorganisms that could infect seeds. The sterilized soils are then tubed using small- and medium-sized polythene bags.

Seedling production. The germination sheds and stand-out beds are prepared in advance to accommodate polythene tubes for germination. The propagation of Araucariaceae is from seeds through direct or broadcast sowing methods. In direct sowing, the seeds are directly sown into the tubes and allowed germination under sheds. Alternatively, the viable seeds are spread on the plastic trays with sterilized soils and allowed for germination under sheds (**Figure 8**). Gunn et al. [17] pointed out that fresh seeds of *Araucaria cunninghamii* take 12–20 days to germinate. The germinated seedlings are then plucked out carefully and transplanted into the polythene tubes. The tubes (with seeds or seedlings) are lightly sprinkled with water to encourage germination.

- *Seedling maintenance.* The following maintenance activities are conducted in the germination sheds (green houses) and stand-out beds for seedling production.

- Weeding—light weeding (0–5 months), medium or high weeding (5 months plus)

- Fertilizer applications—3–4 g of inorganic fertilizers (NPK and boron) are applied to boost plant growth at the juvenile stage.

- Shifting seedlings—after 2–3 months in the germination shed or green house, the seedlings are shifted to stand-out beds for hardening and culling processes.

- Hardening/Culling/Sorting—seedlings are allowed to undergo hardening and culling (**Figure 9**). Dead seedlings are removed, seedlings are graded (A, B, and

C), and slow-growing seedlings (B and C) are separated and treated with fertilizer to boost their growths. Seedlings can be in the stand-out beds for 1–2 years until they reach transplantable size.

- Root pruning—tap roots that protrude out at the base of the polythene bags are pruned to avoid root coiling. Coiled root seedlings when out-planted in the field are subject to fall over (or uprooting) due to shallow penetration of the root systems (shallow rooting-habit) to withstand wind-throw incidents.

- Pest and disease control—a synthetic pesticide (from Blitz chemical group) is applied to control snails that feed on young leaves of seedlings. Pesticides are applied to control insect pests (especially leaf chewing or defoliating insects) and fungi that cause blights and root rots.

- Irrigation—continuous sprinkling of water is done on a daily basis. A good time for watering seedlings is in the mornings (7:00–10:00 AM). Watering is also done in the afternoons, particularly during prolonged dry periods [28].

5.2 Establishment of plantations

The current management of Bulolo and Wau plantations applies 3 × 4 m spacing (12 m^2 per tree) for *Araucaria* spp. and *Agathis robusta* to attain 833 trees per hectare (ha) as initial stocking during establishment (**Figure 10**). Prior to planting,

Figure 10.
Three-year-old second rotation Araucaria cunninghamii crops, Bulolo plantation. Note the line tending done to control spontaneous vegetation, which will shade the young crops.

site preparatory works such as soil mobilization and brushing of vegetation are not usually conducted. Seedlings are immediately planted after a clear-cut operation when the site is cleared from the regrowth of vegetation for reforestation activity. For afforestation, brushing of spontaneous vegetation and control burning are done as site-preparation activities before planting. Most field plantings are planned and executed during rainy seasons to achieve a high survival rate with low mortality. After 4–5 weeks, tree survival counts are conducted for dead trees (mortality) at planted sites and replaced with new seedlings. Mortality count and subsequent replacement are important for maintaining the initial stocking (833 trees/ha) in the compartments.

5.3 Application of silvicultural treatments and harvesting cycle

The main silvicultural treatments applied in plantations after establishments are tending, pruning, application of fertilizer and pesticide, and thinning operations. These treatments are given in order to improve growth performances of the stands and increase crop yield and productivity. The tending operations (clear and line brushing) are scheduled between 1 and 5 years of planting. The *Araucaria* spp. are often self-pruned; however, low pruning can be conducted together with line brushing at an early age. Occasionally, synthetic fertilizer (N, P, K, Mg or B) is applied when trees in certain compartments indicate symptoms of deficiency (yellowing of leaves, stunted growth) of certain macro nutrients. Likewise, pesticide is applied to young trees that show signs of termite (*Coptotermese elisae*) attack as a chemical control measure.

As per the current Bulolo and Wau plantation management decision, moderate- to heavy-intensity thinning activities are scheduled at the age of 5–20 years for the plantations in which non-commercial (thin-to-waste) thinning is conducted (5–7 years old), followed by commercial thinning (17–20 years old) and final harvest (35–40 years old). In thinning operations, theoretically, 50% is removed at each thinning interval, that is, non-commercial and commercial thinning until retained *ca.* 200 stems/ha for final harvest (**Figures 11** and **12**). Basically, low thinning or thinning from below is employed for non-commercial thinning where suppressed and intermediate trees are removed. While for commercial thinning, crown thinning or thinning from above is implemented where the intensity of thinning is expressed by specifying the number of stems or basal area (m^2) per ha to remain after thinning [19, 28]. The present practice in thinning is that any tree that falls within the vicinity of 5–6 m radius from a residual stand is selected and removed. Types of thinning, age and their schedules, average diameter, residual stock, and estimated volume are provided (**Table 3**). The unthinned plots of *Araucaria hunsteinii* and *Agathis rubusta* are shown in **Figures 5** and **13**, respectively. Under the old management of the 1980s, the rotation age of *Araucaria* spp. was 35–40 years; however, the current management decides to allow one commercial thinning only (after non-commercial thinning) and retain the stocking of 200 stems/ha for the final harvest. The non-commercial and commercial thinning treatments create optimum conditions that stimulate prolific growths in residual stocks to reach harvestable size (*ca.* 55 cm diameter-breast-height on average) at 25–30 years (**Figure 14**).

As per the records of CNGT in the 1970s, the mean annual increment (MAI) for *Araucaria* spp. was 20 m^3/ha/year at 40 years of rotation age [19]. According to present Bulolo and Wau plantation management practices, an average of 1.0 m height and 2.0 cm diameter growth increments are attained annually with prudent silvicultural treatments (nursery techniques, fertilization, brushing of vegetation, vine cutting, pruning, thinning) planned and executed on schedule (**Table 3**).

Figure 11.
Commercial thinned compartment with 208 stems/ha as residual stock for final harvest, Bulolo plantation.

Figure 12.
Commercial thinning materials are stored at a landing awaiting log trucks for hauling, Bulolo plantation.

Treatment	Trees/ha	TTR[1]	Age[2]	EAD[3]	FSH[4]	EVH[5]
Non-commercial thin	833	50	5–7	10–15	416	5.522
Commercial thin	416	50	17–20	30–35	208	17.790
Clear cut	208	100	35–40	40–45	Nil	30.206

[1]TTR—Theoretical tree removal (%), [2]Age (years), [3]EAD—Estimated average diameter (cm), [4]FSH—Final stocking (trees/ha), [5]EVH—Estimated volume (m^3/ha).

Table 3.
Thinning schedules, stocking and final harvest of Araucaria spp.

Figure 13.
An unthinned plot of Araucaria hunsteinii at Bulolo University College.

5.4 Final harvest and utilization

As far as harvesting and utilization are concerned, the current plantation management allows an annual allowable cut (AAC) of 63,000 m^3 and 18,000 m^3 from clear-cut (clear-fell) and commercial thinning operations, respectively, from its 12,000 ha forests. These volumes of raw materials (logs) are derived from various compartments of the plantation based on individual compartment history, that

Figure 14.
A compartment of Araucaria cunninghamii due for clear-cut, Bulolo plantation.

is, compartment scheduled for commercial thinning or clear-fell operations. The clear-cut materials are processed into veneer products for plywood manufacturing, while the thinning materials are converted into lumber products. In particular, the *Araucaria hunsteinii* (Klinkii pine) is managed for specific end uses, for example, core veneer in composing plywood and poles for electricity transmission. The *Araucaria* spp. poles and sawn boards are seasoned, machined, and pressure-impregnated with waterborne copper-based preservatives (copper chromium arsenic and alkaline copper quaternary) prior to marketing and utilization [2, 16, 20, 29]. In clear-cut compartments, immediate reforestation with second-cycle *Araucaria* spp. follows suit, while the regrowth of vegetation is still minimal in order to avoid costs involved in heavy brushing and control burning.

5.5 Fire, pests, and disease infestations

The *Araucaria* spp. And *Agathis robusta* are highly susceptible to fire incidents in Bulolo and Wau monoculture plantations. The species are intolerant to fire, with 100% mortality in any fire incidents particularly during the dry season, which falls between the months of June and September per year in PNG. The plantation management is fully equipped with fire-fighting equipment and tools, and the fire-fighting team is always on alert to combat fire. There are three fire lookouts (towers) built at higher locations within the plantations to monitor fire incidents. Any fire incidents are immediately reported to the fire-fighting crews to respond and suppress fire from spreading in the plantations. Huge losses of plantation crops were experienced in 1997 and 1998 El Nino-Southern Oscillation (ENSO) events in PNG [11].

Typical of any monoculture cropping systems, Bulolo and Wau plantations of *Araucaria* spp. are subject to the outbreak of pests and diseases with devastating effects on the crop productivity and economic value. Laufenfels [3] first reported the susceptibility of *Araucaria* spp. from a host of insect pests such as leaf branchlet-mining scolytid, wood-boring weevils, and termites. The major threat to the plantation crops at present is infestations from insect pests, namely *Hylurdrectonus araucariae* Schedl. (Coleoptera: Scolitidae) and subterranean termite (*Coptotermese elisae*) Desneux (Isoptera: Rhinotermitidae). Also, Gray and Lamb [31] reported infestation of the leaves of *Araucaria cunninghamii by H. araucariae*. According to Schneider [32], the larvae of *H. araucariae* bores into the leaves of *A. cunninghamii* and mines the cortex between the healthy green and wilt yellow or brown area of branchlets. The prime targets are 4–11-year-old crops that are confined to compartments located at cooler climatic conditions. The insect pest gradually kills the hosts (*A. cunninghamii*) over a period of time. On the other hand, subterranean termite (*C. elisae*) enters the hosts (*Araucaria* spp.) via injured roots or bark and bore (tunnel) into the cambium and heartwood (**Figure 15**). Once inside the wood, the termite intensifies its boring activity until killing the hosts. The symptoms on the termite-infested trees include yellowing or browning of tips of branchlets, defoliation at the upper crown, and mud galleries constructed at the base of the trees [32]. Often the main targets are young *A. cunninghamii* crops of 4–10 years old, while *A. hunsteinii* and *Agathis robusta* are unaffected by the termite. Usually, young trees growing under stress or weak trees as a result of disease and wound infliction release a volatile allelochemical (kairomone), which is sensed by the termite to initiate its attack [32]. At the moment, the plantation management applies a termiticide (trade name: Termido) as a chemical control

Figure 15.
Subterranean termite (Coptotermes elisae) attack on young Araucaria cunninghamii stem, Bulolo plantation.

measure with limited success. The application of Termido is an expensive exercise, and the control of the subterranean termite is difficult due to the fact that the termite colony lives underground and in most cases are unaffected by the termiticide [32]. A biological control measure undertaken now is the burning of tree residues (debris such as branches, tops, and stumps) after log extraction from clear-fell operations, which become a food source for the termite. The biological control measure looks promising, and further research works are required to substantiate the effectiveness of the technique.

6. Conclusion

The three species *Araucaria hunsteinii, A. cunninghamii*, and *Agathis robusta* of *Araucariaceae* flowers all year round in the PNG's environmental conditions. The seeds of these conifers have low viabilities and should be kept dry (avoid moisture uptake) during extraction from mature cones and refrigerated immediately at recommended temperatures to maintain the quality. The seeds have good dormancy in cool storage as long as they are kept dry and protected from rodents. The conifers are easily propagated from seeds with low mortality in nurseries if proper nursery techniques are applied to raise seedlings for plantation establishment purposes. The conifers have the ability to adapt well and grow prolifically (with 25–30 years cutting cycle) in Bulolo and Wau environmental conditions, except for *Agathis robusta* where its silviculture and cutting cycle are unknown. This opens up research opportunities for silviculture of *A. robusta* in order to promote the species for plantation cultivation. The species of Araucariceae have excellent working characteristics for lumber, treated poles, and plywood products. In PNG, the importance of the species as plantation crops will be maintained with timely application of silvicultural practices. The only threats to plantation of *Araucaria* spp. are fire and termite attack. Burning of forest biomass left behind after log extraction from clear-cut operations is a promising biological control measure for pest (subterranean termite). The global climate change phenomenon will influence the tree physiological processes (photosynthesis, respiration, phenology of flowering and fruiting characteristics) as well as outbreaks of fire, pests, and diseases in the plantations. For instance, climate change potentially affects biotic and abiotic factors, which in turn causes imbalances in macro/micro climatic conditions, ecological functions, soil properties, and forest hydrology that have direct bearing on tree growth and productivity. For sustainable plantation management in the future, one must have a fair understanding of the effects of climate change so that mitigating strategies can be adopted to address the potential impacts to tree crops.

Acknowledgements

I am grateful to the following staff of Bulolo University College: Charles Feriwok and David Tobesa for providing information on nursery techniques, Louis Veisami for forest measurement, Olo Gebia for plant taxonomy, Ryan Dagoro and Samson Aguadi for map work. Also, my appreciation goes to PNGFA's Bulolo plantation officers: Steven Keki, Leo Tohichem, Timothy Sawaraba and Awasa Gebob for providing basic information on seeds and silviculture of PNG's Araucaricaeae. Next, I thank forestry undergraduate Theophyllus Bogan for providing field photographs for the *Araucaria* spp. and *Agathis robusta*.

Author details

Benson Kumuli Gusamo
Forestry Department, Bulolo University College, PNG University of Technology, Papua New Guinea

*Address all correspondence to: benson.gusamo@pnguot.ac.pg

References

[1] Rendle BJ. World Timbers. 3. Asia & Australia & New Zealand. Vol. 3. London: Ernest Benn Ltd; 1970. 175 p

[2] Havel JJ. Forest Botany. Part 2. Botanical Taxonomy. Vol. 3. Training Manual for Forestry College. Port Moresby: Papua New Guinea Department of Forest; 1975. pp. 1-20

[3] de Laufenfels IJ. The external morphology of coniferous leaves. Phytomorphology. 1953;**3**(1,2):1-20. In: Enright, N.J. and Hill, R.S. (eds.) Ecology of the Southern conifers. Smithsonian Institute Press. Washington D.C

[4] Whitmore TC. An Introduction to Tropical Rain Forest. 7th ed. New York: Oxford University Press; 1998. 282 p

[5] Farjon A. A Handbook of the World's Conifers. Vol. 1 & 2. Leiden and Boston: E. J. Brill; 2010. 1112 p

[6] IPNI. International Plant Names Index. The Royal Botanic Gardens, Kew, Harvard University Herbaria & Libraries and Australian National Herbarium. 2023. Available from: http://www.ipni.org [Accessed: July 25, 2023]

[7] Silba J. An International Census of the Coniferae. *Phytologia* Memoir No. 8. Corvallis, OR: H.N. Moldenke and A.L. Moldenke; 1986

[8] Orwa C, Mutua A, Kindt R, Jamnadass R, Anthony S. Agroforestree Database: A Tree Reference and Selection Guide Version 4.0. 2009. Available from: http://www.worldagroforestry.org/sites/treedbs/treedatabases.asp

[9] Reis MS, Ladio A, Peroni N. Landscapes with Araucaria in South America: Evidence of a cultural dimension. Ecology and Society. 2014;**19**(2):43. DOI: 10.5751/ES-06163-190243

[10] Boland DJ, Brooker MIH, Chippendale GM, Hall N, Hyland BPM, Johnston RD, et al. Forest Trees of Australia. Melbourne, Australia: CSIRO; 1985

[11] McKinty MH. Silviculture of Tropical Mixed Forests of Melanesia (Papua New Guinea-Solomon Island-Vanuatu). Department of Forestry. Lae, PNG: The PNG University of Technology; 1999. 90 p

[12] Farjon A. *Agathis borneensis*. The IUCN Red List of Threatened Species. UK; 2013. DOI: 10.2305/IUCN.UK.2013-1.RLTS.T202905A2757743.en

[13] Fowler A, Palmer J, Salinger J, Ogden J. Dendroclimatic interpretation of tree-rings in *Agathis australis* (kauri). 2. Evidence of a significant relationship with ENSO. Journal of the Royal Society of New Zealand. 2000;**30**(3):277-292

[14] de Laubenfels DJ. Agathis montana? Travaux du Laboratoire Forestier de Toulouse. 1969;**8**(5):2

[15] Enright NJ. The Araucaria forests of New Guinea. In: Gressitt JL, editor. Biogeography and Ecology of New Guinea. Monographiae Biologicae Vol. 42. The Hague: Dr. W Junk Publishers; 1982. DOI: 10.1007/978-94-009-8632-9

[16] Arentz F, Keating WG, Ilic J. Araucaria A.L. Juss. In: Soerianegara I, Lemmens RHMJ, editors. Timber Trees: Major Commercial Timbers. Plant Resources of South-East Asia (PROSEA). Vol. 5, no, 1. Leiden: Backhuys Publishers; 1994. pp. 108-114

[17] Gunn B, Agiwa A, Bosimbi D, Brammall B, Jarua L, Uwamariya A. Seed Handling and Propagation of Papua

New Guinea's Tree Species. Canberra, Australia: CSIRO Forestry and Forest Products; 2004. 89 p

[18] PNGFA. Procedures for Exporting Logs. Port Moresby, Papua New Guinea: PNG Forest Authority; 1996. 34 p

[19] Evans J. Tree Planting for Industrial, Social, Environmental, and Agroforestry Purposes. Plantation Forestry in the Tropics. 2nd ed. Oxford, UK: Oxford University Press; 1992. 403 p

[20] Eddowes PJ. Commercial Timbers of Papua New Guinea. Hohola, Papua New Guinea: Forest Products Research Centre; 1977. 195 p

[21] Shearman PL, Bryan JE, Ash J, Hunnam P, Mackey B, Lokes B. The State of the Forests of Papua New Guinea: Mapping the Extent and Condition of Forest Cover and Measuring the Drivers of Forest Change in the Period 1972-2002. Port Moresby, PNG: University of Papua New Guinea; 2008. 148 p

[22] Bryan JE, Shearman PL, Aoro G, Wavine F, Zerry J. The current state of PNG's forests and changes between 2002 & 2014. In: Bryan JE, Shearman PL, editors. The State of the Forests of Papua New Guinea 2014: Measuring Change over the Period 2002-2014. Port Moresby, PNG: University of Papua New Guinea; 2015. 209 p

[23] PNGFA. PNG Forest Authority Corporate Plan 2021-2030. Port Moresby, Papua New Guinea: PNG Forest Authority; 2020. 27 p

[24] The Plant List. Version 1.1. Published on the Internet. 2013. Available from: http://www.theplantlist.org/ [Accessed: August 18, 2023]

[25] Blackmore S, Tootill E. The Penguin Dictionary of Botany. London: Penguin Books; 1984. 391 p

[26] Enright NJ, Hill RS. Araucariaceae; life history and ecology. In: Enright NJ, Hill RS, editors. Ecology of the Southern Conifers. Washington D.C.: Smithsonian Institute Press; 1995

[27] Whitmore TC. Utilization, potential and conservation of *Agathis*, a genus of tropical Asian conifers. Economic Botany. 1980;**34**(1):1-12. Available from: http://www.jstor.org/stable/4254132

[28] Romijn KD. Plantation Silviculture. Training Manual for the PNG Forestry College. Vol. 18. Hohola, PNG: PNG Forestry College, Department of Forest; 1986. pp. 1-278

[29] Havel JJ. Plantation establishment of klinki pine (*Araucaria hunsteinii*) in New Guinea. Commonwealth Forestry Review. 1965;**44**:172-186

[30] Iverson KR, van der Mel M, Chambers PC, Emerson LHS, et al. The Economic Development of the Territory of Papua and New Guinea. Report of a Mission Organized by the International Bank for Reconstruction and Development at the Request of the Government of the Commonwealth of Australia. Baltimore, Maryland: The Johns Hopkins Press; 1965

[31] Gray B, Lamb KP. Biology of *Hylurdrectonus araucariae* Schedl. (Coleoptera: Scolytidae), a pest of hoop pine plantations in New Guinea. Bulletin of Entomological Research. 1975;**65**:21-32

[32] Schneider MF. Entomology. A Textbook for Students, Agriculturists and Foresters in Papua New Guinea. Training Manual No. 19. Papua New Guinea: Bulolo University College; 1999. 312 p

Chapter 3

Exploring the Biometric Traits and Potential of Radiata Pine (*Pinus radiata* D. Don) as a Non-Native Species for Sustainable Forest Systems in Portugal

Teresa Fidalgo Fonseca, Renato N.M. Costa, Carlos Pacheco Marques, José Luis Louzada and Ana Cristina Gonçalves

Abstract

This chapter aims to provide information on biometric traits of radiata pine (*Pinus radiata* D. Don) outside its natural range, considering as a case study the use of the species in Portugal. The specific objectives of this study are: i) characterising the species; ii) its management; iii) its provisioning potential. To achieve the latter, data on the biometric characteristics of radiata pine trees in Portugal was compiled and analyzed. Briefly, the approach followed employs an equation developed to predict the stem volume of individual trees, which is then coupled with the inherent wood basic density to provide oven-dried biomass estimates. The volume equation demonstrated a noticeable goodness-of-fit (R^2 = 0.994 and standard error of the residuals = 0.026 m^3) across the entire range of diameters within the dataset (ranging from 7.5 to 45 cm). Additionally, a proposed wood density value of 460 kg/m^3 is put forth as a representative value for the species. The tree stem biomass (and sequestered carbon) is then generalized to the stand unit. The results show that the species compares favourably with maritime pine in terms of wood provisioning and usage, broadening the options of pine species to consider in Portugal for reforestation or afforestation programs.

Keywords: Monterey pine, distribution, silviculture, volume, biomass, wood technology

1. Introduction

Provision is one of the categories of benefits provided by forests, with timber and fiber being the most common products in this category [1]. Given the current challenges faced by the timber industry and the impact of climate change, it is crucial to plan for the selection of appropriate forest species for afforestation and reforestation. Radiata pine, also known by the common names Monterey pine and insignis pine, has been widely used in timber plantations in many temperate regions due to its fast growth and wood properties. In Portugal, the interest in the species was translated into forestation programs, carried out in the 20th century, in different locations around the country, to study its adaptability. The interest was recently renewed in view of the threats of climate change and the pressure to obtain coniferous woody material for the Portuguese industry. The decrease of maritime pine (*Pinus pinaster* Ait.) forest area, in the last decades [2], heavily affected by fires and pests (the most expressive being nematode), and the expectations of changes in its distribution area due to climate change have motivated the use of other pine species. Among these, the radiata pine is of particular interest to the industry, namely due to the remarkable growth reported for the species in countries with a long tradition of managing the species, such as Chile [3].

To make an informed decision about utilizing this species, it is crucial to collect information about its characteristics and assess its potential for use in the wood supply service beyond its native range. This chapter provides information on radiata pine and being structured into four major parts. The first part covers the species' distribution, biotic and abiotic disturbances, and the diversity and sustainability of its forest systems. The second part summarizes information on the management of the species, including silvicultural systems and practices. The third part focuses on the utilization of the species for provisioning, with a specific emphasis on its application in Portugal. Specifically, the case study delves into three aspects: (i) assessing tree stem volume using volume equations; (ii) presenting information on the species' wood properties; and (iii) generalizing the volume quantification for biomass assessment. This study employs biometric data obtained from felled *Pinus radiata* trees, encompassing measurements of tree diameter, height, volume, and bark thickness. This dataset was originally compiled and utilized in prior studies conducted by one of the authors [4, 5], serving as a foundational resource for formulating a volume equation tailored to this particular species. Subsequently, samples of wood cores were acquired from standing trees to facilitate an analysis of wood characteristics and contribute to biomass assessment.

2. Distribution and ecology of *Pinus radiata*

Pinus radiata D. Don is an evergreen conifer that belongs to the *Pinaceae* family and *Pinus* genus. This species has several common names, namely radiata pine, Monterey pine, and insignis pine [6]. Its natural distribution is in California [6], in the coastal zone (**Figure 1**) from the parallel 35°30' N to the parallel 37° N, corresponding to an area of about 200 km long, 10 km wide, and an elevation less or equal to 300 m [8]. The actual area of distribution is about 67,020 ha, in 5 patches, four inland with a total area of 58,270 ha (with patches of 13,450, 31,002, 6358, and 7461 ha, respectively, from north to south) and one in the Guadalupe island (*circa* 8750 ha) [7]. This species in their natural range from a conservation point of view is considered vulnerable and endangered status [6].

Radiata pine has been introduced in Australia, New Zealand, Spain, Argentina, Chile, Uruguay, Kenya, South Africa, Portugal [5, 8], United Kingdom, France [9], and Italy [10]. The largest areas are found in New Zealand, Chile, and Australia [11].

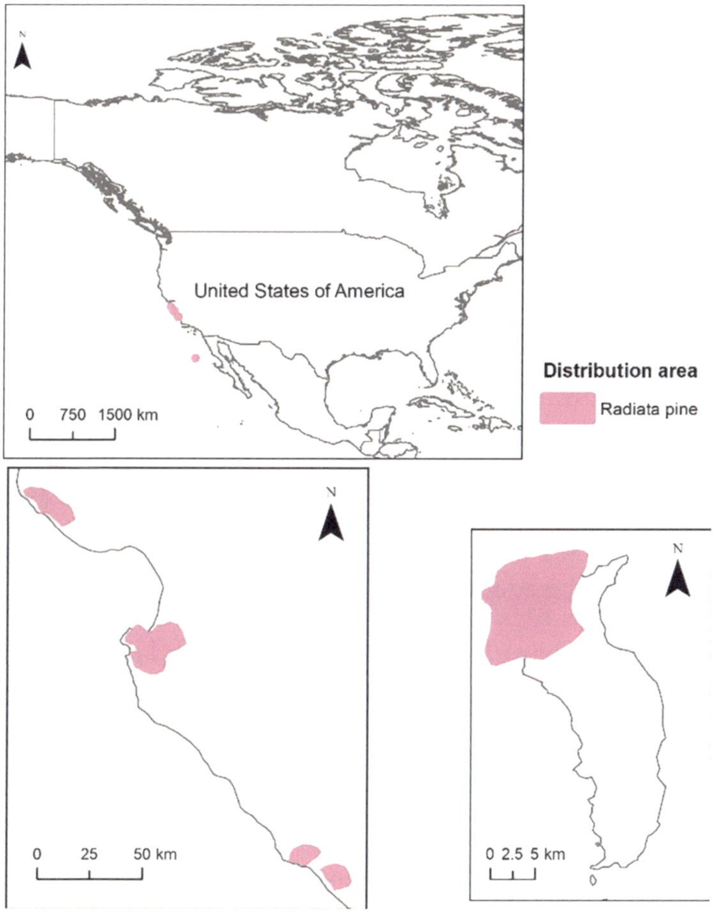

Figure 1.
Natural distribution of radiata pine (data source: [7]).

From these countries, the largest areas of plantations are located in New Zealand and Australia [6], with a total area of about 4 million hectares.

In its area of natural distribution stands of radiata pine can be pure or mixed with several conifers and/or broadleaved species [8]. *Pinus radiata* has a relatively short lifespan (*circa* 50 years), with a diameter at breast height reaching 60–120 m and a height of 30–50 m. Its crown presents high variability from conical with strong epinastic control and strong annual height growth (that can reach 1–2 m) to a flat and shorter length [10, 11]. Its branch longevity is large, resulting in larger crowns when compared to *Pinus pinaster*. Natural pruning is not frequent, and branches, live and dead, remain in the tree for many years, thus artificial pruning is prescribed to produce knot-free stems. In general, the stems are straight, but have some tendency to fork, due likely to pest attacks; and to develop a curvature in the stem, which results in reaction wood and thus wood with less interesting technological properties [11]. It is a species semi-shade tolerant though it develops better in full sunlight [8]. Its root system is superficial, sometimes without a pivot root, but well developed laterally up to a soil depth of 60 cm [10, 11], most of which up to 30 cm of soil depth [11]. Fruiting begins at 7–8 years old and becomes abundant at 15–20 years old. Fruit-full development is reached in the second year, cones are serotinous, opening after hot weather or fire, usually in the spring after fruit-full development, and have numerous seeds [6, 10, 11]. The cones after hot weather or fire open their scales releasing part of their seeds and close the scales shortly after. As cones have a high number of viable seeds and are maintained in the trees for many years, it enables several seed rains, which favor its regeneration [11]. It is a very fast-growing species [9].

This species develops in a wide range of temperatures from −5–41°C, with a mean annual temperature between 12 and 14°C, mean annual temperatures of 9–11°C in the winter and 16–18°C in the summer [8, 11]. Prefers mean annual precipitations between 380 mm and 890 mm, concentrated in the winter (an average of 300–510 mm) and less than 50 mm in the remaining months [8]. The lack of precipitation in July and August is balanced by mist precipitation, due to high air relative humidity (about 60–70%) and fogs [8, 10]. Radiata pine resistance to strong maritime winds is good [8, 11], but is sensitive to frost [10, 11]. It prefers deep well-drained soils and tolerates acidic soils but is sensitive to high zinc content [8], heavy and compact soils as well as soils with drainage problems [11].

3. Silviculture and management

In its natural range of distribution low-density stands and gaps in the canopy created by silvicultural practices have, in general, high number of seedlings and saplings. This regeneration is originated by the multiple seed rains, which are the result of the species traits. The cones are kept in the tree for many years. They open the scales after hot weather or fire releasing the seeds and closing the scales shortly after, which allows the trees to storage enough seeds for the regeneration of the stands. Moreover, the semi-shade tolerance during the young stages of development enables the seedlings and saplings to survive and grow under the canopy in semi-light environments. For example, it can live under the canopy of *Quercus agrifolia* for many years. Furthermore, due to the fast growth rates of the regeneration (during the first 15 years), it can outcompete other species [11].

Outside its natural range, most stands result from artificial regeneration (frequently plantation), resulting in pure even-aged stands, at regular spacing and with plantation densities ranging from 815 stems/ha to 1666 stems/ha and rotations between 25 and 35 years [11, 12].

For Portugal, pure even-aged stands with a plantation initial density of 1666 trees/ha with short rotation (25–30 years) and 2 or 3 thinnings from below at 8–10 years old, 15–18 years old, and *circa* 20 years old, respectively for the first, second and third thinning were recommended. Natural pruning occurs at higher densities, yet at low densities, it may be needed. Control of spontaneous vegetation is required. The trees react to fertilization with an increase in growth. It has some sensitivity to processionary attacks, which might decrease its growth [12].

For Spain, two models of silviculture have been found for Galicia [11]. One model considers the plantation with an initial density of 833 stems/ha and a rotation of 25 years. The other model considers the plantation with an initial density of 1142 stems/ha and a rotation of 35 years. In both models three thinnings are prescribed, the first non-commercial and the latter two commercial. The latter two are thinning from below, the intensity moderate to heavy and the periodicity of about 5 years. Two prunings are considered at the first and second thinning. It also considers the control of spontaneous vegetation with a periodicity of about 5–6 years [11].

Silvicultural management depends on the site quality and the management aims. In Chile [3], for high-productivity sites, it is considered an initial density of 1000 to 1100 trees/ha and 25 years of rotation length. There are three pruning operations in a subset of 500–600 trees (2.1 m—5 yr., 3.6 m—6 yr., and 5.5 m—7 yr) and two thinning operations (at 5 and 9 yr), aiming to reduce density to around 500 trees/ha. In low-productivity sites, the initial density is similar (1000 to 1100 trees/ha), with a lower rotation length (21 to 24 years). There is no prescription for intervention, other than a phytosanitary pruning. The final density is around 800 trees/ha assuming an expected average mortality of 20% of the trees.

4. Wood traits, tree volume, and biomass

Understanding the variability of the wood characteristics of a species not only allows the understanding of the development conditions of the trees but also to evaluate the quality of the wood produced by it and to infer its most appropriate applications. Wood density is a straightforward measure of the amount of woody material in a given volume. It can be quickly and accurately determined, and is often highly heritable with significant variability, making it an ideal target for genetic modification. Additionally, wood density is closely linked to numerous important properties and technological features that are essential to the production and utilization of forest products. As such, it is the most informative index for understanding the fundamental characteristics of wood [13–17].

Wood density can be expressed in multiple ways, including anhydrous density, saturated density, density at 12% moisture, and basic density, with the latter being one of the most referred to in the literature. Expressed by the ratio between the anhydrous weight of the wood and its saturated volume, it indicates the amount of mass present in a given volume of wood without considering the presence of water. Although it does not represent any real situation from the wood utilization point of view, it is of great value in studying the woody variation of trees. Besides being relatively fast and accurate, its determination is perfectly feasible in irregularly shaped and small samples, requiring minimal equipment. Basic wood density also has the added advantage of providing information that can be used to estimate the biomass of wood in dry weight, given that its volume is known (e.g., [18, 19]).

To determine the volume of a tree, indirect methods are commonly used. These methods involve equations based on the allometric model or other mathematical relationships, which typically estimate stem volume (v) based on the diameter measured at 1.30 m above ground level (d) or that diameter combined with the total height of the tree (h) (see, [20, 21], for further information on this topic). Biomass can be evaluated by destructive methods or estimated by equations that are similar to those used for modeling volume. For the species *Pinus radiata*, a number of volume and biomass models were proposed (see, [22–28]). For an overview of available models for the species, with reference to the location they were developed for, see the GlobAllomeTree database (http://www.globallometree.org/; [29]). Biomass can also be estimated by combining volume estimates with wood density, as mentioned. This method is especially useful when stem volume data or a predictive volume model is available and the goal is to obtain wood biomass estimates from volume. From the search carried out, and to the best knowledge of the authors, there are no studies on *Pinus radiata* that follow this approach.

5. Assessing wood traits, tree volume, and biomass: a case study in Portugal

5.1 Material and methods

The case study joins information on individual tree characteristics collected in seven forest stands of radiate pine across mainland Portugal (Odemira, Montejunto, Lavos, Leirosa, Aveiro, Marão and Cabreira) and processed in previous studies authored by the co-author R. Costa [4, 5], with additional measurements taken in 2022 and 2023, in tree additional sites (Moimenta da Beira, Vila Real and Malcata). The former data support the study's analysis of the tree's volume while the second data collection was specifically intended for targeting the evaluation of wood characteristics in radiate pine trees. **Figure 2** shows the localization of the mentioned sampled radiate pine in Portugal used in the case study.

Figure 3 provides an overview of the materials, activities, and techniques used in the case study to develop tree volume equations, assess wood characteristics, and estimate the stem biomass of radiate pine. The combination of both methods and data sets allowed for the non-destructive estimate of the biomass of the species' stem which can be extended to quantifying the carbon stored in this component. Yet, auxiliary

Figure 2.
Locations in mainland Portugal where radiate pine data was collected.

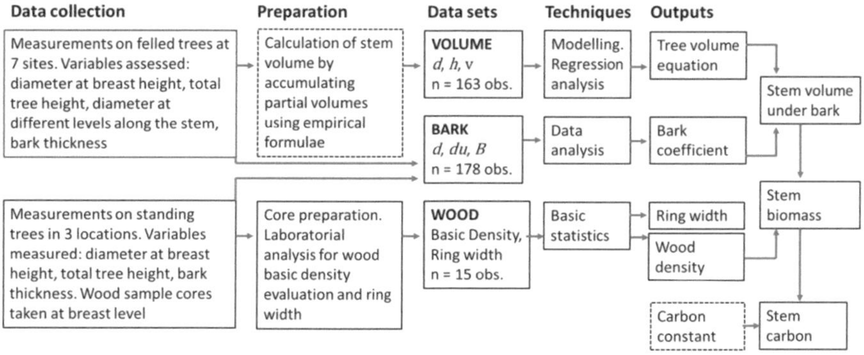

Figure 3.
Schematic diagram of material and methodological preparation and development techniques used in the study.

calculations might be needed. When determining the volume of trees using volume equations, the volume usually includes the stem and the bark. To estimate the volume of wood, the portion of the volume that corresponds to the bark must be excluded. To do this, bark thickness is assessed in a sample set of trees, followed by the assessment of the bark reduction factor (k) for the species. This factor is then combined with information on the volume of the stem with bark to estimate the volume of wood in the stem (v_u). This methodology was proposed in Meyer [30] and can be found in forest measurement literature (e.g., [19, 31]). The main procedures used for the case study are described, including data characterization to assess the bark reduction factor and equations to quantify k and v_u.

This section is structured into separate subsections to favor the presentation of the distinct data sets and methodologies followed. Subsequent subsections encapsulate both the presentation and discussion of the results.

5.1.1 Volume equation development

Pinus radiata tree data from seven locations in mainland Portugal available from previous studies and processed by one of the co-authors [4, 5] served as a foundation for formulating a volume equation tailored to this particular tree species. The data was obtained from 120 trees that were cut down during thinning from below performed in five sites (Odemira, Montejunto, Lavos, Leirosa, Aveiro,) and from 43 standing trees in sites where felling was not possible (Marão and Cabreira). The sites are shown in **Figure 2**.

The data collected encompassed measurements of tree diameter at breast height over bark (d, cm) and total tree height (h, m). The stems of the felled trees were measured at 0.1 m and 1.30 m above ground for diameter and over 2 m length sections along the stem. The diameters were measured with a caliper with cross measurements and the length with a tape. The volume of each section was then computed using analytical formulae (e.g., the cylinder formula for the portion corresponding to the stump, the cone formula for the upper part of the trunk, and Smalian's formula for the intermediate logs). The volume of the stem (v, m^3) was obtained by summing the volumes of all portions from the base up to stem height (h) in accordance with the rigorous cubage method [19]. The diameter variable was measured to the nearest mm and the length of the sections was recorded to the nearest cm. The volume of standing trees was assessed using a Bitterlich relascope, by identifying the formal height and applying the Pressler-Bitterlich cubage method [5, 19, 32].

Table 1 provides a summary of the dataset, while **Figure 4** displays scatter plots depicting the relationship between tree diameter at breast height over bark and total tree height with stem volume for the VOLUME data set.

Upon analyzing the observations, tests were conducted to produce a volume equation for the species. The tests included common formulations of volume equations found in literature, such as the ones mentioned in ([33], p. 8), or in other reference books (e.g., [19] or [20]). The considered potential regressors were diameter and height, which were found to be clearly associated with volume, as shown in **Figure 4**. To address potential heteroscedasticity, the logarithm of the volume was utilized as the response variable instead of using the original variable volume. The exogenous variables tree diameter and height, as well as their transformations and interactions, were tested. The models were fitted with ordinary least squares (OLS) regression.

Statistical analysis was conducted to evaluate the goodness-of-fit of the estimated models using the coefficient of determination (R^2), and the Root Mean Square Error (*RMSE*) to identify departures to the ordinary least squares assumptions. The Spearman's rank correlation test was applied to detect the presence of heteroscedasticity and the variance inflation factor (*VIF*) was evaluated to assess multicollinearity [34, 35]. The accuracy of the models was also assessed using the Furnival index (*FI*) [36], regarded as an average standard deviation of the residuals transformed into units of volume. The model development was performed using the JMP® software (JMP®, Version 17.2.0. SAS Institute Inc., Cary, NC, 1989–2023), and the model among the ones essayed with the best performance, based on the fitting criteria (R^2, *RMSE*, and *FI* statistics) and OLS assumptions were ultimately chosen. The chosen fitted model was adjusted by incorporating the correction term (exp(MSE/2)) suggested by [37] to account for the transformation bias resulting from transforming the logarithm of volume values back into volume estimates as a final step.

Variable	Min	Average	Max	Standard deviation	CV (%)
d (cm)	7.5	23.2	44.9	11.4	49
h (m)	6.5	17.5	28.5	6.1	35
v (m^3)	0.012	0.481	1.723	0.481	100

Table 1.
Characterization of the data set VOLUME (n = 163 *obs.).*

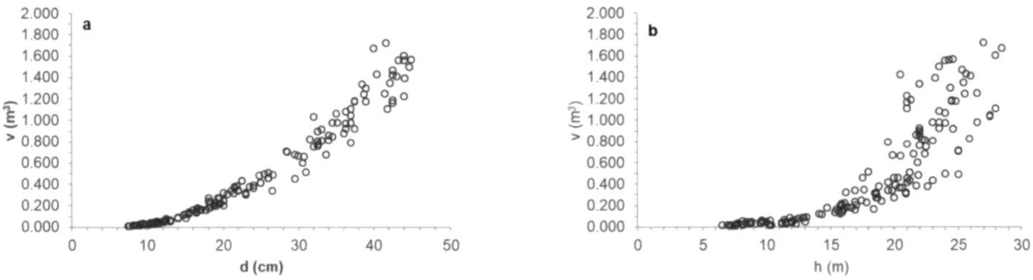

Figure 4.
Graphical representation of the observations used in the volume modeling (n = 163 *obs.). Scatter plots of stem volume in relation to the variables diameter at breast height (a) and tree total height (b).*

5.1.2 Wood basic density and radial growth assessment

The material used to study the basic density and radial growth (width of growth rings) of radiata pine refers to samples of wood cores collected in three stands located in Moimenta da Beira, Vila Real, and Malcata (**Figure 2**), with ages of approximately 12, 40, and 50 years, respectively. At each of these sites, a selection of 3–6 radiata pines was made. The selection process involved choosing trees encompassing the observed diameter classes within a 5 cm range, in each stand, with upright, healthy stems, and excluding those that displayed possible formation of reaction wood.

A radial sample from cambium to pith was extracted from each tree with an increment borer (Pressler auger) about 5 mm thick, taken at 1.30 m above ground level and as perpendicular as possible to the tree vertical axis. The diameter over bark at breast height and total height values of the sampled trees are provided in **Table 2**.

Each core sample properly identified was prepared for laboratory evaluation, with placement in a support structure followed by sanding. The width of each growth ring was measured (using a magnifying glass with a micrometric displacement table with an accuracy of 0.001 mm) and then sectioned into specimens consisting of three growth rings. The basic density was calculated by the ratio between the anhydrous weight and the saturated volume of each wood specimen. The saturated volume of the specimens was determined by the method of impulsion in distilled water of the previously saturated samples (Archimedes' principle), with the aid of an analytical balance with an accuracy of ± 0.0001 g. The anhydrous weight was quantified after the specimens were placed in an oven at 100 ± 3°C until constant weight stabilization. The analyses were carried out at the Forest Products Laboratory of the University of Trás-os-Montes and Alto Douro.

5.1.3 Bark thickness, bark reduction coefficient, and stem volume under bark evaluations

The thickness of the bark (B) was evaluated on the sample radiata pine trees mentioned in subsections 5.1.1 and 5.1.2, for a total of 178 trees, by measuring two opposite sides of the trunk with a thickness gauge that has an accuracy of 1 mm. Measuring the double thickness of the bark ($2B$) allows the diameter under bark (du) to be evaluated from the diameter over bark (d) using the expression: $du = d - 2B$.

Table 3 presents a summary of the diameter with and without bark and bark thickness values measured at 1.30 m (variables d, du, and B, respectively), which constitutes the BARK subset of data (**Figure 3**).

From the information on d and du, the bark reduction factor (k) is determined. This factor is calculated following Meyer [30] as the ratio between the sum of the diameters of the stem with bark and without bark (Eq. (1)):

$$k = \frac{\sum_{i=1}^{n} d_u}{\sum_{i=1}^{n} d} \tag{1}$$

Variable	Min	Average	Max	Standard deviation	CV (%)
d (cm)	13.7	28.8	50.0	11.6	40
h (m)	10.0	16.3	21.5	3.6	22

Table 2.
Characterization of the data set WOOD (n = 15 obs.).

Variable	Min	Average	Max	Standard deviation	CV (%)
d (cm)	7.5	17.7	50.0	9.4	53
du (cm)	7.0	16.0	45.0	8.1	51
2B (cm)	0.2	1.7	6.4	1.5	87

Table 3.
Characterization of the data set BARK (n = 178 obs.).

According to the same reference [30], the volume of the stem without bark can be approximated by the relationship shown in Eq. (2), where k is the bark reduction factor calculated through Eq. (1).

$$v_u = k^2 v \tag{2}$$

5.2 Results and discussion

5.2.1 Stem volume over and under bark

After the fitting procedures on the VOLUME subset of data, the model that has shown overall goodness-of-fit statistics is the equation that considers as an explanatory variable the product of the square diameter and height (d^2h). Results of the estimation are provided in **Table 4**.

Analysis of the residuals did not show departures to normality. Additionally, as the model considers a single regressor, there is no multicollinearity (VIF = 1). The Spearman's rank correlation test did not evidence the presence of heteroscedasticity ($\rho = -0.1182$, $p\text{-}value = 0.1328$). After some adjustments including applying the Baskerville factor, $\exp\left(0.013^2/2\right)$, to the selected model, Eq. (3) is obtained. This equation can be utilized to approximate the total stem volume over the bark of radiata pine in Portugal.

$$v = 3.6242 \times 10^{-5} d^{1.9856} h^{0.9928} \tag{3}$$

The proposed equation (Eq. (3)) for stem volume estimation over bark, adequately describes the volume pattern observed for the range of diameter and height values measured, as shown in **Figure 5** (continuous line). The graph in **Figure 5** depicts a trend line (dotted line) for volume values derived from the volume equation utilized for the maritime pine species observed in Portugal's national forest inventory, NFI6 [2]. The trend line is provided for comparison and assumes that both pines have a similar height-diameter relationship, which needs to be tested. If this assumption holds true, it can be inferred that radiata pine produces lower volume in the stem than maritime pine, although it tends to compare favorably with the latter, namely for trees up to the 30 cm diameter class (calculated mean difference of estimated values circa 0.020 m^3, based on n = 108 obs.).

Model	β_0 (s_{β_0})	β_1 (s_{β_1})	R^2	RMSE	FI (m^3)	FI (%)
$\ln v = \beta_0 + \beta_1 \ln(d^2h)$	−10.2318 (0.0558)	0.9928 (0.0063)	0.994	0.013	0.026	5.4

Table 4.
Parameters of the selected volume equation and goodness-of-fit statistics (n = 163 obs.).

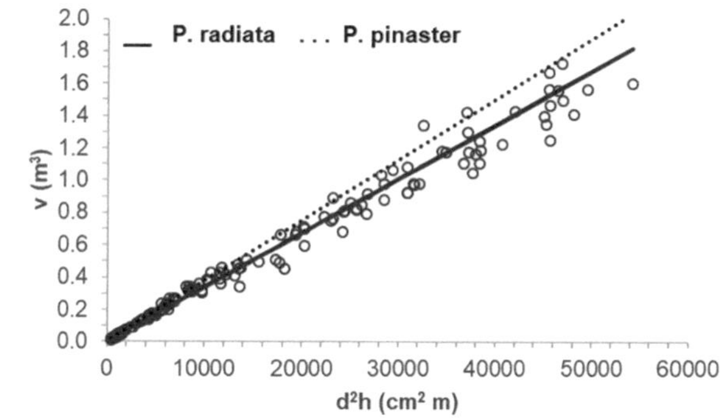

Figure 5.
Graphical representation of the data set (o) and the fitted line for volume estimation of radiata pine over bark (Eq. (3), continuous line). Estimates of stem volume over bark for maritime pine (NFI6, [2] dotted line).

Determining the volume of a tree stem without bark (v_u) based on its volume with bark (v) can be easily accomplished by using the bark reduction factor (k). Estimation of the bark reduction factor for radiata pine through Eq. (1) yield, k = 0.9055, around 0.91.

To estimate the volume of the stem without bark (v_u), the bark reduction factor (k) is applied in Eq. (3), as previously shown (Eq. (2)). Thus, the equation proposed for stem volume estimation, excluding bark, is:

$$v_u = 2.972 \times 10^{-5} d^{1.9856} h^{0.9928} \quad (4)$$

Considering the bark coefficient, the estimated value of k (a. 0.91) is within the range of values mentioned in [31] (0.87–0.4) for this variable, which can vary with the species, the age, and site factors. Marques et al. ([21] and cited references herein) point out limitations with the expedited methods (Eqs. 1 and 2) for determining bark amount and volume under the bark. These methods assume a constant bark reduction factor along the trunk, and the linearity may not apply to all diameters. Additionally, bark thickness vary due to factors such as growing conditions and tree age/size. The case study did not explicitly analyze the impact of growing conditions and age. However, the variation was assessed based on tree size. It was found that there was a slight decrease in k values from the smallest to the largest diameter classes (0.93–0.90). This study's broader geographical scope, the large sample size used, and the short range of variation provide confidence in the method's reliability and supports the use of an average value of k = 0.91.

For maritime pine, Duarte [38] presents values of the bark coefficient ranging from 0.65 to 0.95, with average values around 0.81. Based on the sample data used in the study, Duarte [38] mentions that the average proportion of bark in comparison to the total volume of the stem is 30%. When comparing the bark coefficient values of maritime pine reported by Duarte [38] with the values obtained in the case study for radiata pine (k *circa* 0.81 and 0.91, respectively), radiata pine has a greater wood volume and less bark than maritime pine for the same amount of stem volume over bark.

5.2.2 Wood density

Table 5 shows the values of basic density and ring width, per tree, in each of the three sampled sites (Moimenta da Beira, Vila Real, and Malcata). From the values, it

	Basic density (kg/m³)			Ring width (mm)		
Tree	**Moimenta da Beira**	**V. Real**	**Malcata**	**Moimenta da Beira**	**V. Real**	**Malcata**
1	369	513	431	11.0	3.6	3.1
2	350	464	438	10.0	4.4	2.2
3	365	453	539	9.7	1.8	2.7
4		461	444		2.0	1.9
5		493	503		3.5	1.7
6		436	434		2.7	3.0
Average	361	470	465	10.2	3.0	2.4
Standard deviation	10.0	28.1	45.2	0.7	1.0	0.6
CV %	2.8	6.0	9.7	6.7	33.7	24.1

Table 5.
Values of basic density and ring width, by site and tree.

can be observed that *Pinus radiata* trees in Moimenta da Beira have an average wood density of 361 kg/m^3. The average value per tree ranges from 350 to 369 kg/m^3. These trees are still young, with only 8 to 9 rings at 1.30 m level, and are mainly composed of juvenile wood, which explains the relatively low wood density.

In contrast, the trees in Vila Real and Malcata stands are much older (around 40 to 50 years old) and are composed of both juvenile and adult wood, yielding an average wood density value of 470 kg/m^3. The average value per tree ranges from 431 to 539 kg/m^3. These values are similar to those obtained for *Pinus pinaster* wood by Fonseca and Lousada [39] in three stands in the north of Portugal aged between 35 to 55 years old, which presented an average value of 489 kg/m^3 (ranging from 383 to 528 kg/m^3 between trees), as well as by Louzada [40] in two stands around 80 years old in Gerês (426 kg/m^3) and Marinha Grande (479 kg/m^3).

As *P. radiata* revealed wood density values very similar to *P. pinaster* (the most used coniferous species in Portugal), we can therefore conclude that these two species produce wood with very identical characteristics. Hence, whenever necessary, *P. radiata* can be used as a substitute for *P. pinaster*, being suitable for use in carpentry, framing, furniture, veneer, laminate, and chipboard, as well as with identical carbon-sequestration capacity.

Regarding the ring width, the trees in Moimenta da Beira have a considerably higher radial growth (ring width) than the other two sites (Moimenta da Beira = 10.2 mm; Vila Real = 3.0 mm and Malcata = 2.4 mm) since they are exclusively composed of juvenile wood. In comparison, Fonseca and Lousada [39] reports an average value of 3.9 mm for ring width in adult *P. pinaster* trees, while Louzada [40] reports 2.5 and 1.9 mm for the stands in Gerês and Marinha Grande, respectively.

5.2.3 Biomass and carbon

Estimating the biomass of the stem is straightforward when an equation is available that estimates the volume of the wood (Eq. (4)), complemented by the value of its basic density. Based on the analyses performed for the case study, the recommended value for wood basic density is 460 kg/m^3.

The biomass of wood in dry weight (b, kg) is calculated by multiplying the volume of wood (v_u, m^3) by the average wood basic density. In the alternative, the basic density value is applied directly in Eq. (4). Thus, the equation proposed for dry biomass estimation of stem wood, is:

$$b = 0.01367d^{1.9856}h^{0.9928} \tag{5}$$

To the best of the authors' knowledge, Eq. (5) is the first model that is available for estimating the stem biomass of *Pinus radiata* trees in Portugal. The method used involved combining the stem volume data with wood density to convert volume measurements into mass measurements. This approach has not been previously applied to the studied species, according to the literature. Allometric equations have been proposed to estimate radiata pine biomass (e.g. [22–24, 27, 28]). However, these equations were developed using data from other regions and might result in biased estimations in Portugal.

Montero et al. [24] found that 78.5% of the dry matter in this particular species is located in the above-ground portion of the tree with 83% of this dry matter in the stem (including the bark but free of branches). These figures allow to consider that roughly 2/3 of the biomass of the tree is in the stem.

For evaluating the carbon content in this component, a conversion factor of 0.5 (or a more specific value if known) can be used, assuming that 50% of the dry biomass is composed of carbon. To estimate the carbon stored in the stem component, the estimates are adjusted using a coefficient that takes into account the ratio between the molecular weight of carbon dioxide and the atomic weight of carbon (44/12). This calculation gives the amount of carbon stored in CO_2 equivalent (CO_2e) (Eq. (6)).

$$CO2e = 0.02506d^{1.9856}h^{0.9928} \tag{6}$$

The relationships derived from tree variables and wood parameters (Eqs. (3) to (5)) form the basis to assess the supply capacity of species in afforestation programs, filling a national knowledge gap.

6. Conclusions

The results obtained in the case study permit the estimation of tree stem volume and dry biomass of radiata pine, which is of the utmost importance for characterizing the wood provisioning service potential and carbon quantification for the species. The information complements previous studies about radiata pine in Portugal and enables meaningful comparisons with other pine species, thereby supporting informed decision-making in forest and timber management planning. The results of this study reveal that radiata pine exhibits wood properties and offers provisioning services that are comparable to those of maritime pine. These findings widen the range of pine species that can be considered for national reforestation or afforestation programs.

When making decisions about which species to use, it is important to consider edapho-climatic factors, as they have a significant impact on the success and productivity of the selected species. To refine decision-making, future analyses must include a comparative study of the growth patterns of various pine species under identical conditions. This comprehensive understanding will enable more informed decisions in forestry management, ensuring that the chosen species can successfully adapt to their specific environments.

Acknowledgements

Thanks are due to the International Union of Forest Research Organizations (IUFRO), namely Division 1 (Silviculture), unit 1.01.10 Ecology and Silviculture of Pine, and Centro PINUS association, for promoting fruitful discussions on the silviculture and management of pine forests that have contributed to the decision of writing this document. Carlos Fernandes from CIFAP-UTAD is thanked for his valuable contribution to data collection for the WOOD dataset.

Funding

This work was partially funded by the FCT – Fundação para a Ciência e a Tecnologia) to Forest Research Centre (CEF), within project UIDB/00239/2020, to Center for the Research and Technology of Agro-Environmental and Biological Sciences (CITAB), under project UIDB/04033/2020, and to MED – Mediterranean Institute for Agriculture, Environment and Development under the Project UIDB/05183/2020.

Conflict of interest

The authors declare no conflict of interest.

Author details

Teresa Fidalgo Fonseca[1,2*], Renato N.M. Costa[3,4], Carlos Pacheco Marques[1], José Luis Louzada[1,5] and Ana Cristina Gonçalves[6]

1 Department of Forestry Sciences and Landscape Architecture (CIFAP), University of Trás-os-Montes and Alto Douro, Vila Real, Portugal

2 Forest Research Centre (CEF), School of Agriculture University of Lisbon, Lisboa, Portugal

3 Institute for Nature Conservation and Forests (ICNF), Lisbon, Portugal

4 Union of Portuguese Speaking Capital Cities (UCCLA), Lisbon, Portugal

5 Centre for the Research and Technology of Agro-Environmental and Biological Sciences (CITAB), University of Trás-os-Montes and Alto Douro, Vila Real, Portugal

6 MED – Mediterranean Institute for Agriculture, Environment and Development & CHANGE – Global Change and Sustainability Institute, Institute of Research and Advanced Education (IIFA), Department of Rural Engineering, University of Évora, Évora, Portugal

*Address all correspondence to: tfonseca@utad.pt

References

[1] FAO. Ecosystem Services & Biodiversity (ESB) [Internet]. Food and Agriculture Organization of the United Nations. Available from: http://www.fao.org/ecosystem-services-biodiversity/en/ [Acceessed: October 3, 2023]

[2] ICNF. 6° Inventário Florestal Nacional: 2015 Relatório Final [Internet]. Lisboa: ICNF; 2019. 284 p. Available from: https://www.icnf.pt/api/file/doc/c8cc40b3b7ec8541 [Accessed: August 2, 2023]

[3] Ahumada R. Webinar Management of Monterey Pine (*Pinus radiata*) in Chile. International Series Pine Silviculture: International Success Factors [Internet]. Available from: https://www.youtube.com/watch?v=cer_l3pBcoI&ab_channel=CentroPINUS [Accessed: October 3, 2023]

[4] Costa R. Técnicas Dendrométricas. Lisboa: Instituto Superior de Agronomia, Universidade Técnica de Lisboa; 1970. p. 11

[5] Costa R. Comparação entre métodos e instrumentos na cubagem do volume das árvores em pé: caso da espécie *Pinus radiata* D. Don. Lisboa: Direcção-Geral de Fomento Florestal; 1980. 12 p. (Estudos)

[6] Christopher J. Earle. Conifers. The Gymnosperm Database [Internet]. Conifers. Available from: https://www.conifers.org/topics/index.php; [Accessed: July 30, 2023]

[7] Atlas US Trees. Atlas of United States Trees. Digit. Represent. Tree Species Range Maps. Washington: United States Forest Service; [Internet]. Available from: https://web.archive.org/web/20170127093428/;https://gec.cr.usgs.gov/data/little/ [Accessed: May 8, 2023]

[8] Correia A, Oliveira ÂC. Principais espécies florestais com interesse para Portugal, Zonas de influência atlântica. Lisboa: Direcção-Geral das Florestas; 2003. 187 p. (Estudos e Informação)

[9] Lousã M. Outros pinheiros. In: Silva JS, editor. Público, Comunicação SocialÁrvores e florestas de Portugal Pinhais e eucalipais A Floresta Cultivada. SA: Fundação Luso-Americana para o desenvolvimento; 2007. pp. 133-146

[10] Costa R. Contribuição para a definição de áreas ecologicamente favoráveis à espécie *Pinus radiata* D. Don em Portugal. Lisboa: Fundo de Fomento Florestal; 1976. 39 p. (Estudos)

[11] Castro A, Rodríguez M, Alboreca A, Rodríguez F. Manual de selvicultura del Pino Radiata en Galicia. Galicia: Agrobyte; 2023

[12] Góis E. A floresta portuguesa: sua importância e descrição das espécies de maior interesse. Lisboa: Portucel; 1991

[13] Panshin AJ, de Zeeuw C. Textbook of wood technology. In: Structure, Identification, Uses, and Properties of the Commercial Woods of the United States. 3rd ed. Vol. I. USA: Vaux, H.J: McGraw Hill Book Company; 1970. 705 p. (The American Forestry Series)

[14] Haygreen JG, Bowyer JL. Forest Products and Wood Science: An Introduction. Ames, IOWA: Iowa State University Press; 1982. 520 p

[15] Zobel BJ, Van Buijtenen JP. In: Timell TE, editor. Wood Variation: Its Causes and Control. Berlin: Springer-Verlag; 1989. 363 p

[16] Tsoumis G. Science and Technology of Wood: Structure, Properties, Utilization. Vol. 115. New York: Van

Nostrand Reinhold New York; 1991. 494 p

[17] Zobel BJ, Jett JB. In: Timell TE, editor. Genetics of Wood Production. Berlin: Springer-Verlag; 1995. 337 p. (Springer Series in Wood Science)

[18] Picard N, Saint-Andre L, Henry M. Manual for Building Tree Volume and Biomass Allometric Equations: From Field Measurement to Prediction. Rome: FAO and Centre de Coopération Internationale en Recherche Agronomique pour le Développement; 2012. 215 p

[19] Marques C, Fonseca T, Duarte J. Guia Prático de Avaliações Florestais—Dendrometria. Faro, Portugal: Sílabas & Desafios; 2017. 230 p

[20] Burkhart HE, Tomé M. Modeling Forest Trees and Stands [Internet]. Dordrecht: Springer Netherlands; 2012. 457 p. Available from: https://link.springer.com/10.1007/978-90-481-3170-9 [Accessed: October 2, 2023]

[21] Marques C, Fonseca T, Duarte J. Guia Prático de Avaliações Florestais—Inventário Florestal e Modelação Estatística. Faro, Portugal: Sílabas & Desafios; 2018. 302 p

[22] Wang X, Bi H, Ximenes F, Ramos J, Li Y. Product and residue biomass equations for individual trees in rotation age *Pinus radiata* stands under three thinning regimes in New South Wales, Australia. Forests. 2017;**8**(11):439

[23] Cartes-Rodríguez E, Rubilar-Pons R, Acuña-Carmona E, Cancino-Cancino J, Rodríguez-Toro J, Burgos-Tornería Y. Potential of *Pinus radiata* plantations for use of harvest residues in characteristic soils of South-Central Chile. Revista Chapingo serie ciencias forestales y del ambiente. 2016;**22**(2):221-233

[24] Montero G, Ruiz-Peinado R, Munoz M. Producción de biomasa y fijación de CO_2 por los bosques españoles. Vol. 13. Madrid: INIA-Instituto Nacional de Investigación y Tecnología Agraria y Alimentaria; 2005 (Coléccion Monografías INIA: Serie Florestal; 270 p

[25] Zianis D, Muukkonen P, Mäkipää R, Mencuccini M. Biomass and stem volume equations for tree species in Europe. Silva Fenn, Monogr. 2005; **2005**(4):1-63

[26] Rubilar RA, Allen HL, Alvarez JS, Albaugh TJ, Fox TR, Stape JL. Silvicultural manipulation and site effect on above and belowground biomass equations for young *Pinus radiata*. Biomass and Bioenergy. 2010;**34**(12): 1825-1837

[27] Instituto Florestal Chile I. Tablas de volumen para Pino insigne (*Pinus radiata* D. Don.) en Chile. Vol. 2. Chile: Instituto Forestal; 1965. 14 p. (Boletín técnico;

[28] Instituto Florestal Chile I. Tablas de volumen para Pino insigne (*Pinus radiata* D. Don.) en Chile. Vol. 1. Chile: Instituto Forestal; 1962. 8 p. (Boletín técnico)

[29] GlobAllomeTree-Assessing volume, biomass and carbon stocks of trees and forests [Internet]. Available from: https://www.fao.org/sustainable-forest-management/toolbox/tools/tool-detail/en/c/217974/ [Accessed: October 3, 2023]

[30] Meyer HA. Bark volume determination in trees. Journal of Forestry. 1946;**44**(12):1067-1070

[31] Husch B, Beers TW, JAK J. Forest Mensuration. 4th ed. USA: John Wiley & Sons; 2003. 443 p

[32] Bell JF, Bitterlich W, Iles K, Ruthner G. Spiegel Relaskop® Manual.

Austria: Silvanus Forstbedarf GmbH; 2018. 21 p

[33] Clutter JL, Forston JC, Brister GH, Bailey RL. Timber Management: A Quantitative Approach. USA: John Wiley & Sons; 1983. 333 p

[34] Myers RH. Classical and Modern Regression with Applications. 2nd ed. USA: PWS Publishers; 1986. 488 p

[35] Neter J, Kutner MH, Nachtsheim CJ, Wasserman W. Applied Linear Regression Models. 3rd ed. USA: Irwin; 1996. 720 p

[36] Furnival GM. An index for comparing equations used in constructing volume tables. Forest Science. 1961;7(4):337-341

[37] Baskerville GL. Use of logarithmic regression in the estimation of plant biomass. Canadian Journal of Forest Research. 1972;2(1):49-53

[38] Duarte JPC. Estudos biométricos em Pinheiro bravo: Configuração do perfil do tronco, volumes e percentagem de casca [thesis]. Portugal: Universidade de Trás-os-Montes e Alto Douro; 2001

[39] Fonseca FM, Lousada JLPC. Variação na madeira de Pinus pinaster Ait: o comprimento e as dimensões transversais das fibras: a densidade, o crescimento ea qualidade físico-mecânica da madeira. Vila Real, Portugal: Universidade de Trás-os-Montes e Alto Douro; 2000. 242 p. (Série Técnico-Científica, Ciências Aplicadas)

[40] Louzada J. Variação fenotípica e genética em características estruturais na madeira de *Pinus pinaster* Ait. O comprimento das fibras e a densidade até aos 80 anos de idade das árvores. Parâmetros genéticos na evolução juvenil —adulto das componentes da densidade da madeira. Vol. 143. Vila Real, Portugal: Universidade de Trás-os-Montes e Alto Douro; 2000. 293 p. (Série Didáctica, Ciências Aplicadas

Chapter 4

Influence of Silvicultural Operations on the Growth and Wood Density Properties of Mediterranean Pines

Daniel Moreno-Fernández, Andrea Hevia, Iciar Alberdi and Isabel Cañellas

Abstract

Silvicultural operations are widely used for forest regeneration and promotion of tree growth by reducing competition. The main aim of pruning, on the other hand, is to disrupt vertical fuel continuity and enhance wood quality, although the impact of silviculture on wood properties has scarcely been studied in the case of Mediterranean conifer forests. Our main goal is to synthesize the primary findings regarding the impact of thinning and pruning on tree growth and wood density of Mediterranean conifers. For this purpose, we used data from three thinning and pruning trials in Central Spain, specifically in forests of *Pinus sylvestris* and two subspecies of *Pinus nigra*. Our results indicate that thinning enhanced tree growth for the three species but did not significantly affect wood density. In contrast, no significant effects of pruning were observed, either on tree growth or on wood density. We concluded that thinning in combination with pruning is a suitable way to promote tree growth without compromising wood quality.

Keywords: knot-free timber, microdensitometry, Mediterranean forestry, sustainable forest management, timer quality

1. Introduction

High-quality wood, such as sawn wood and veneer, typically necessitates high-grade logs, large in diameter, containing mostly clear wood, with straight stems and a significant amount of heartwood [1]. For this, it is desirable that any knots are limited to a narrow central core in order to obtain the so-called clear wood [2], free of knots, of more valuable wood [3]. However, clear wood not only increases the quality of the highest-value by-products by removing visual defects but also reduces the influence of knots on the magnitude of pith eccentricity, stem curvature, and bending [1, 4, 5].

Besides the abovementioned wood properties that are desirable for high-quality uses, other characteristics such as wood density are a major physical criterion for wood quality [6] since they are related to many other aspects of quality such as wood strength and shrinkage, fiber properties, and flexibility or stiffness, among others [7, 8]. Furthermore, it has been demonstrated that wood density affects carbon storage [9–11].

The variability of wood density is not only dependent on the functional group or species [12, 13] but has also been observed to vary among provenances [14] and climate [15], a high level of variability being attributable to this factor. Other factors, such as site conditions or genetics, also affect wood density [16]. In addition, wood density does not remain constant across the trunk but varies both in radial (from pith to bark) and axial directions, forming juvenile and adult wood (also known as corewood and outerwood, respectively) [17, 18]. Furthermore, variations in wood density also occur at the intra-ring level because of the differentiation between early and latewood. Earlywood usually presents lower density values than latewood, although its section is normally larger [19]. In contrast to earlywood, the density and section of latewood increase with cambial age [20, 21]. In this regard, the proportion of latewood emerged as a key variable in the characterization of wood density.

Silvicultural treatments, such as thinning, have commonly been employed to reduce stand density and increase the diameter growth rates of the remaining trees [22, 23] having a major influence on wood quality [8]. Pruning operations, meanwhile, involve the removal of branches, which contributes to limiting knots and other branch-related defects to a central "knotty core" [13] resulting in more valuable wood (clear wood) [3]. Additionally, both operations can play a key role in the crown fire hazard [24] since thinning can reduce fuel loading and connectivity and pruning disrupts the vertical continuity of fuel, reducing the severity of forest fires [25]. On the other hand, pruning usually has a negative impact on tree diameter growth [3, 26, 27] although the magnitude of this effect depends on the proportion of crown removed by pruning [4, 28]. As regards the impact of silvicultural operations on wood density, most studies state that thinning has a limited impact on wood density [9, 20, 29, 30]. As for pruning, while [16] and [31] reported an increment in wood density after pruning, other authors such as [32] found no significant influence of pruning on wood density properties.

The way in which silviculture affects wood density is an important issue as higher growth rates coupled with lower wood density as a result of thinning and/or pruning operations could lead to bias in the estimation of forest biomass and therefore carbon accounting [33, 34].

The joint impact of thinning and pruning on tree growth and wood density has been poorly evaluated in Mediterranean conifer forests. In this study, our objective is to synthesize and evaluate the primary findings presented by [2, 35] regarding the impact of these silvicultural practices on tree growth and wood density. For this purpose, we used data from three thinning and pruning trials in Central Spain, specifically in forests of *Pinus sylvestris L.* and two subspecies of *Pinus nigra* Arnold. These species are important not only for the forestry sector in Spain but also from an ecological perspective.

2. Material and methods

2.1 Material

We used data from three thinning and pruning trials located in Central Spain. These experimental trials were established in monospecific reforestations of *P. sylvestris*, *Pinus*

Feature	*P. sylvestris*	*P. nigra*
Coordinates	40°520 N, 3°510 W	41°020 N, 3°040 W
Altitude (m asl)	1650	1050
Aspect	North facing	None
Slope (%)	10–40	0–3
Average annual rainfall (mm)	1062	620
Average annual temperature (°C)	7	10.5

Table 1.
Ecological characteristics of the thinning trials.

nigra subsp. *nigra* Arnold, and *Pinus nigra* subsp. *salzmannii* (Dunal) Franco. The trials were initiated at the beginning of the 1990s when the *P. sylvestris* stand was 37 years old, the *P. nigra nigra* 26 years old, and the *P. nigra salzmannii* 31. While both the *P. nigra* trials are adjacent and share similar ecological characteristics, the *P. sylvestris* stand is located at a higher altitude with colder and wetter conditions (see **Table 1**).

The *P. sylvestris* trial was initiated in 1991 when the first thinning was undertaken. Since then, five inventories have been conducted, in 1991, 1996, 2001, 2006, and 2011, including diameter at breast height measurements. The experiment consisted of nine permanent plots, each covering 0.1 hectares, with a 10 m buffer area to eliminate the edge effect (**Table 2**). Three treatments were applied (i.e., three plots per treatment): control treatment in which only dead trees were felled (C), thinning from below without pruning (T), and thinning from below combined with pruning (TP). Thinning intensity was around 30% in terms of basal area. In the TP treatment, trees were pruned to a height of 6 meters, and 40 dominant and codominant trees per plot were selected for pruning. This resulted in a stand density of 400 trees per hectare, ensuring an adequate number of pruned trees to achieve the desired stand density during the regeneration phase (200–300 trees per hectare).

In 2001, the second thinning operation (ca. 15% in basal area) was carried out in the study plots. However, it is important to note that one plot per treatment had to be excluded from the analysis due to a severe storm in January 1996, which resulted in significant snow-throws.

The *Pinus nigra nigra* experiment with six plots was established in 1993. Dasometric inventories were conducted in 1993, 1998, 2002, 2006, and 2011 (**Table 1**). **Figure 1** illustrates the appearance of a thinned plot in May 2023.

In the case of *Pinus nigra salzmannii*, the establishment, inventories, and thinnings followed the same schedule as the *Pinus nigra nigra* trial. The same three treatments (C, T, and TP), with two replicates in six plots (0.1 ha), were evaluated in the *P. nigra salzmannii* trial and in six plots of *Pinus nigra nigra*. Thinning intensity, however, was greater for *P. nigra nigra* (40% in terms of basal area) than for *Pinus nigra salzmannii* (25–30%) In 2006, the second thinning with an intensity of 16% of the basal area was performed in these two trials.

To investigate the impact of silvicultural operations on wood density, six cores were taken at a height of 1.3 m above ground level (breast height) from six trees per treatment and taxa in January 2013. Therefore, the full dataset included 18 cores for taxa, i.e., a total of 54 cores. These trees were selected from the second and third quartiles of the diametric distribution, representing the codominant trees within the stand. It is worth noting that both dominant and codominant trees were present in the stand

Treatment	First thinning					Second thinning				
	Age	N	Dg	BA	%BA	Age	N	Dg	BA	%BA
Pinus sylvestris										
C	37	2332	14.5	38.5	0.7*	47	1997	17.2	46.4	9.9*
T	37	2037	14.6	34.1	28.2	47	928	20.7	31.2	18.6
TP	37	2082	14.4	33.9	34.8	47	821	21.2	29.0	14.4
Pinus nigra nigra										
C	26	1392	18.2	36.2	0.0	39	1250	21.2	44.1	0.0
T	26	1447	17.8	36.0	41.9	39	725	23.6	31.7	17.6
TP	26	1455	17.7	35.8	40.2	39	756	23.3	32.2	16.5
Pinus nigra salzmannii										
C	31	1597	15.7	30.9	0.0	44	1446	18.1	37.2	0.0
T	31	1574	15.6	30.1	24.8	44	1064	19.6	32.1	16.9
TP	31	1498	16.6	32.4	30.4	44	907	20.9	31.1	16.9

*C = control treatment. T = thinning without pruning. TP = thinning with pruning the best trees. Adapted from [35]. *Natural mortality.*

Table 2.
Mean values of quadratic mean diameter (dg; cm) before thinning, basal area (BA; m^2 ha^{-1}) before thinning and percentage of basal area removed (%BA) per treatment and thinning intensity.

Figure 1.
Photograph of a P. nigra nigra thinned plot in May 2023 (author D. Moreno-Fernández).

until the start of the regeneration period. To obtain the wood cores, a 5 mm diameter increment borer was used. In the TP treatment, increment cores were exclusively taken from the pruned trees, as these were the focus of this particular treatment.

2.2 X-ray microdensitometry measurements

In the laboratory, each increment core obtained was mounted on a wooden holder. The cores were then cut into longitudinal radial strips, approximately 1 mm thick, using a twin-blade saw. To remove resins, the samples were refluxed in 96% ethanol using a Soxhlet apparatus. The refluxing process lasted 24 hours for *P. sylvestris* and 48 hours for *P. nigra*. The resulting thin strips were then stored under constant temperature and humidity conditions before being subjected to X-ray analysis. X-ray imaging was performed using an Itrax Multiscanner (Cox Analytical Systems, Mölndal, Sweden) at the CETEMAS laboratory in Asturias, Spain. The Multiscanner, equipped with a Cu-tube operating at 30 KV, 50 mA, 25 ms with 20 μm steps, produced radiographic images that were later analyzed using WinDendro software (Regent Instruments, Québec, QC, Canada). From the radiographic images, average wood density values for tree rings (*RD*, in g cm^{-3}) and the proportion of latewood density relative to the entire ring width (*LWP*, in %) were extracted. This extraction involved calibrating the greyscale intensities to wood densities using a light calibration curve derived from a calibration wedge [36]. Cross-dating accuracy was assessed using statistical parameters provided by the dendrochronological software COFECHA (University of Arizona, Tucson, AZ, USA) [37].

2.3 Statistical analyses

To evaluate the effect of the silvicultural operations on the stand variables and tree wood density of each taxon, we used linear mixed models. The assumption is made that measurements obtained from the same tree or plot exhibit a stronger correlation than those from different trees or plots. Additionally, measurements taken in closer proximity in time on the same tree or plot are expected to have a higher degree of correlation than those taken further apart in time [38, 39]. Consequently, the traditional assumptions of independent and homogeneous error variance are no longer applicable due to the inherent correlation pattern among observations. To solve this, we entered an autocorrelative structure of errors and random effects in the wood density models [39, 40]. As response variables, we considered the *id*, tree diameter increment (mm $year^{-1}$) calculated as the ratio of the difference between two consecutive forest inventories and the temporal lapse between inventories, *RD*, ring density (g cm^{-3}), and *LWP*, percentage of latewood (%). Thus, the linear mixed model included an intercept, treatment, Time (periods between inventories for *id* and year for *RD* and *LWP*), as well as their interaction, as fixed effects. The model also included random intercept effects for the tree and plot to account for the abovementioned correlations. Both random effects follow a normal distribution with mean zero and variance σ_b^2 and σ_s^2. We included the diameter recorded in the previous inventory and the ratio of the basal area of larger trees to the plot basal area (*BAL/B*) in the *id* model to control the effects attributable to tree size and competition [41, 42]. In the dataset for *P. sylvetris*, there were 1533 *id* observations for C, 696 for T and 630 for TP. Meanwhile, the dataset for *P. nigra* contained 1025 *id* observations for C, 549 for T and 576 for TP, while the dataset for *P. nigra salzmannii* included 1188 *id* observations for C, 818 for T and 693 for TP.

To account for the initial differences in wood density properties, a covariate (*AM5*) was used. This covariate was calculated as the arithmetic mean of the specific wood property within the annual rings formed 5 years prior to the commencement of the trials [29, 43, 44]. Finally, the models included a ε random error term. All the statistical analyses were run in R using the "lme" function of the "nlme" package [45] and the restricted maximum likelihood option. Model structures were compared using Akaike's Information Criterion. We used Tukey's post hoc test to conduct pairwise comparisons between group means to identify which groups differ significantly from each other using the "emmeans" package.

3. Results

3.1 Impact of silvicultural operations on diameter increment

The species displaying the largest tree diameter increment, regardless of the thinning treatment applied, were *P. nigra* and *P. sylvestris* (mean growth for both species was 2.8 mm $year^{-1}$ with standard deviation of 1.6 mm $year^{-1}$), while *P. nigra salzmannii* exhibited slightly lower growth rates (2.1 ± 1.3 mm $year^{-1}$).

We found a significant effect ($p \leq 0.05$) of the Treatment and Time on tree growth for the three taxa, while the interaction between them was found to be significant in the *P. nigra nigra* trial (**Table 3**). Tukey's post hoc test revealed that the trees in C plots exhibited significantly lower growth than those in thinned plots (**Figure 2**). As regards the thinning treatments, we only found significant differences between T and TP for the *P. nigra* trial, with TP presenting larger diameter increment than T.

The diameter at the beginning of the period and the competition index BAL/B were found to have a significant effect on tree growth (**Table 3**) in both the *P. sylvestris* and the *P. nigra* trials. The diameter at the beginning of the period displayed a

Species	Treatment	Time	Treatment*Time	dbeg	BAL/B
P. sylvestris	<0.0001	<0.0001	n.s.	<0.0001 (−)	<0.0001 (−)
Pinus nigra nigra	<0.0001	<0.05	<0.001	<0.001 (−)	<0.0001 (−)
P. nigra salzmannii	<0.05	<0.0001	n.s.	<0.0001 (−)	<0.0001 (−)

Adapted from [2]. dbeg: diameter at the beginning of each period, respectively, and BAL/B: the ratio between the basal area of the largest trees and the stand basal area. n.s. = non-significant (p > 0.05).

Table 3.
Mixed model results for the diameter increment models for Pinus sylvestris (2859 observations), Pinus nigra nigra (2510 observations), and Pinus nigra salzmannii (2699 observations).

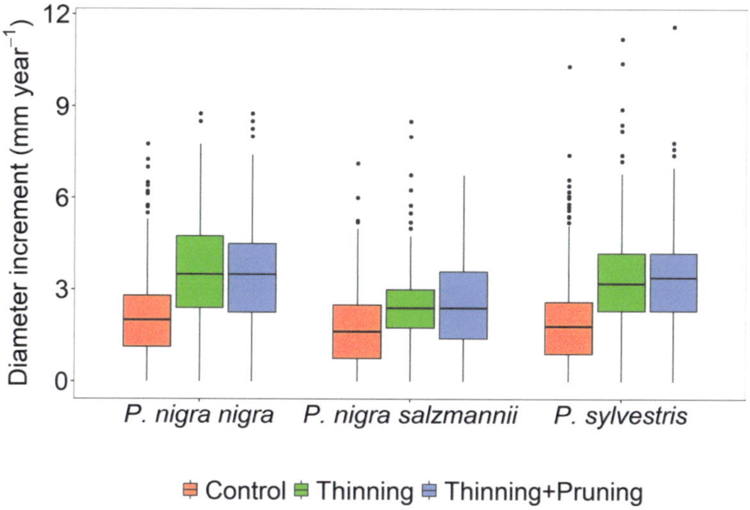

Figure 2.
Boxplot of tree diameter increment for the three studied species and treatments.

negative relationship with diameter increment, indicating that thinner trees exhibited more growth than larger trees. The *BAL/B* was negatively correlated with diameter growth in *P. sylvestris* and both *P. nigra* subpsecies. A negative estimation coefficient for BAL/B implies that trees with larger BAL/B (thinner diameters and more competition) exhibited less growth compared to larger trees. This divergence between the effect of the diameter at the beginning of the period and BAL/B can be explained by the strong, negative correlation between the two variables (α = 99.9%).

3.2 Impact of the silvicultural operations on wood properties

The mean value of *P. nigra salzmannii* wood density was 0.66 kg cm^{-3} with a standard deviation of 0.10 kg cm^{-3}. *P. nigra nigra* a had a wood density mean value of 0.62 ± 0.11 kg cm^{-3}, whereas *P. sylvestris* exhibited a mean value of 0.54 ± 0.08 kg cm^{-3}. As regards latewood percentage, this variable reached a value of 37.5 ± 14.4% for *P. nigra salzmannii*, 31.9 ± 10.7% for *P. nigra nigra*, and 28.0 ± 11.2% for *P. sylvestris*. All of these values were calculated using the temporal series starting from experiment initiation (1991 for *P. sylvestris* and 1993 for *P. nigra*) up until the date the cores were extracted.

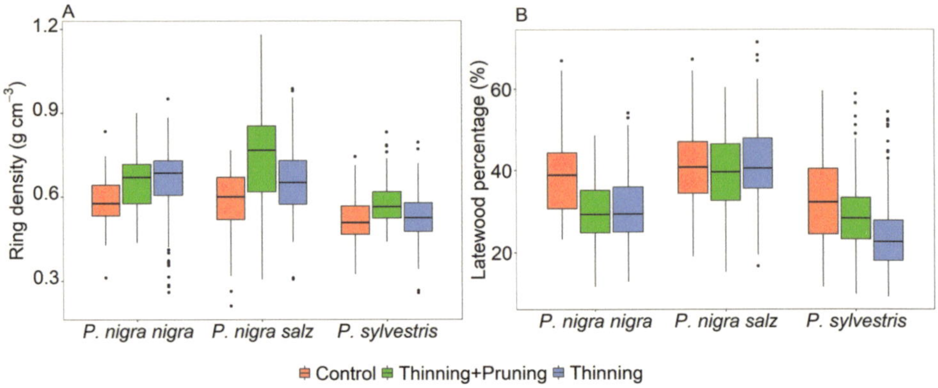

Figure 3.
Boxplot of (A) tree ring density values and (B) latewood percentage for the three studied species and treatments. The data used ranged from experiment initiation to core collection date.

Variable	AM5	Time	Treatment	Treatment x Time
Pinus sylvestris				
RD	<0.0001	n.s.	n.s.	n.s.
LWP	<0.05	n.s.	n.s.	n.s.
Pinus nigra nigra				
RD	<0.0001	n.s.	n.s.	n.s.
LWP	n.s.	n.s.	n.s.	n.s.
Pinus nigra salzmannii				
RD	<0.0001	n.s.	n.s.	n.s.
LWP	n.s.	n.s.	n.s.	n.s.

Adapted from [35]. n.s. = non-significant ($p \geq 0.05$).

Table 4.
P-value of AM5 (5-year arithmetic mean prior to the initiation of the trials), time, treatment, and interaction of treatment × time in the RD (ring density) and LWP (percentage of latewood) models.

Despite the visual differences observed among treatments as shown in **Figure 3**, there was no statistically significant effect of the treatment factor on the two wood density variables studied (RD and LWP) for any of the three species. Additionally, the covariate *AM5* appeared to be significant in all models except for *LWP* in both *P. nigra nigra* and *P. nigra salzmannii* (**Table 4**), suggesting that the abovementioned differences between treatments may be partially associated with RD and LWP temporal trends prior to initiating the trials.

4. Discussion

In this chapter, we have evaluated the impacts of common silvicultural treatments on tree growth and wood properties (wood density and LWP) of two dominant pine species found in the Spanish mountains.

This positive effect of thinnings on tree diameter growth is in agreement with the findings of most previous studies [46–48]. In addition to promoting secondary growth in trees, thinning may enhance components of tree resilience (*sensu* [49]) during drought periods [50] serving as a climate change adaptation tool. These effects are not accompanied by a significant loss in wood quality in terms of wood density and latewood proportion, which is in line with previous results reported for conifers [9, 20, 29, 30]. In contrast, [22] reviewed the impacts of thinning on the set of properties defining wood quality in *P. sylvestris* and reported a negative impact of this silvicultural operation. However, many of the properties covered by these authors, such as strength, stiffness, knottiness, distortion, wood heterogeneity, and compression wood, have not been considered here. Moreover, the thinning experiments discussed in [22] were conducted with the future crop trees in mind, aiming to foster the growth of the highest-quality trees. This approach may lead to a more substantial release of space compared to our study, potentially exerting a greater influence on wood quality. Our findings indicate that pruning has a negligible effect on the growth, ring density, and latewood percentage in *P. sylvestris* and *P. nigra* subspecies. This suggests that pruning is an appropriate treatment to remove branches and obtain knot-free timber without a reduction in wood density. Previous studies, however, postulated that pruning significantly impacts tree growth and that this effect is directly related to the percentage of green crown removed [3, 26–28]. It is important to note that we have not quantified the percentage of crown removed during the pruning operations, but both intensity and timing of the pruning and thinning operations were within the schedules of regular forest prescriptions, that is, 6 m pruning in low-size trunks, ca. 15–25 cm wide [51]. Therefore, it is possible that the 6-m pruning treatment eliminated dead branches and the lower part of the crown, which is expected to have low photosynthetic activity. In particular, this would be the expectation in the case of *P. sylvestris*, which is a self-pruning species.

Our results open a window for further research regarding the combination of thinning and pruning: (i) the impact on growth and wood density at different trunk heights and (ii) the effect on other wood quality properties (e.g., strength, stiffness, knottiness, distortion, wood heterogeneity, and compression wood). Additionally, although more information has become available in recent years on the influence of climate and other site conditions on wood density [52–54], the effects of interaction between climate and management or land-use legacies on wood properties are still scarcely understood [34].

5. Conclusion

Our findings provide strong evidence supporting the efficacy of implementing combined silvicultural practices, that is, thinning and 6 m pruning, in Mediterranean middle-aged pine forests. The thinning intensity and pruning height assessed in this study align with established practices in Mediterranean pine forests. Consequently, the findings presented in this chapter offer valuable scientific insights for forest managers, aiding them in their decision-making for the typical forest operations they undertake. It has been evidenced that not only do these silvicultural interventions enhance wood-quality characteristics, such as promoting larger diameters and knot-free timber, but also the wood density remains at the same levels as untreated plots.

Acknowledgements

This work was funded by IFN-2021 (Monitorización de la red de parcelas permanentes de Gestión Forestal y Tratamientos Selvícolas del CIFOR-INIA) and 101056907-PathFinder (The contribution of forest management to climate action: pathways, tradeoffs and co-benefits). We also thank Adam Collins for revising the English grammar as well as the reviewers and editor for their critical input.

Funding

A. Hevia was supported by "Action 7: Grants for the temporary incorporation of postdoctoral research staff, from the Operational Plan for Research Support of the University of Jaén (POAI-UJA)", "Project LITHOFOR, RTI2018-095345-B-C21, Spanish Ministry of Science, Innovation and Universities, R&D Program Oriented to the Challenges of Society, 2018 Call" and by "the fellowship II.4 from VII-PPITUS 2022 (Univ. Sevilla)".

Conflict of interest

The authors declare no conflict of interest.

Author details

Daniel Moreno-Fernández[1*], Andrea Hevia[2,3], Iciar Alberdi[1] and Isabel Cañellas[1]

1 Institute of Forest Sciences (INIA-CSIC), Madrid, Spain

2 Universidad de Jaén, Campus Las Lagunillas s/n, Jaén, Spain

3 Departamento de Biología Vegetal y Ecología, Universidad de Sevilla, Avda. Reina Mercedes s/n, Sevilla, Spain

*Address all correspondence to: daniel.moreno@inia.csic.es

References

[1] Ruano A, Alberdi I, Adame P, Moreno-Fernández D, Amiano AC, Fernández-Golfín J, et al. Improving stem quality assessment based on national forest inventory data: An approach applied to Spanish forests. Annals of Forest Science. 2023;**80**:20. DOI: 10.1186/s13595-023-01187-7

[2] Moreno-Fernández D, Sánchez-González M, Álvarez-González JG, Hevia A, Majada JP, Cañellas I, et al. Response to the interaction of thinning and pruning of pine species in Mediterranean mountains. European Journal of Forest Research. 2014;**133**:833-843. DOI: 10.1007/s10342-014-0800-z

[3] Hevia A, Gabriel Álvarez-González J, Majada J. Comparison of pruning effects on tree growth, productivity and dominance of two major timber conifer species. Forest Ecology and Management. 2016;**374**:82-92. DOI: 10.1016/j.foreco.2016.05.001

[4] Neilsen WA, Pinkard EA. Effects of green pruning on growth of *Pinus radiata*. Canadian Journal of Forest Research. 2003;**33**:2067-2073. DOI: 10.1139/x03-131

[5] Víquez E, Pérez D. Effect of pruning on tree growth, yield, and wood properties of *Tectona grandis* plantations in Costa Rica. Silva Fennica. 2005;**39**:381-390

[6] Alves A, Hevia A, Simões R, Majada J, Alia R, Rodrigues J. Improving spatial synchronization between X-ray and near-infrared spectra information to predict wood density profiles. Wood Science and Technology. 2020;**54**:1151-1164. DOI: 10.1007/s00226-020-01207-z

[7] Erdene-Ochir T, Ishiguri F, Nezu I, Tumenjargal B, Baasan B, Chultem G, et al. Modeling of radial variations of wood properties in naturally regenerated trees of *Betula platyphylla* grown in Selenge, Mongolia. Journal of Wood Science. 2021;**67**:1-10. DOI: 10.1186/s10086-021-01993-5

[8] Macdonald E, Huber J. A review of the effects of silviculture on timber quality of Sitka spruce. Forestry. 2002;**75**:107-138. DOI: 10.1093/forestry/75.2.107

[9] Peltola H, Kilpeläinen A, Sauvala K, Räisänen T, Pekka IV. Effects of early thinning regime and tree status on the radial growth and wood density of scots pine. Silva Fennica. 2007;**41**:489-505. DOI: 10.14214/sf.285

[10] Larjavaara M, Muller-Landau HC. Rethinking the value of high wood density. Functional Ecology. 2010;**24**:701-705. DOI: 10.1111/j.1365-2435.2010.01698.x

[11] Zobel BJ, van Buijtenen JP. Wood Variation. Its Causes and Control. New York: Springer-Verlag; 1989. p. 363

[12] Farias HLS, Pequeno PACL, Silva WR, Melo VF, LCS DC, De Perdiz RO, et al. Amazon forest biomass: Intra- and interspecific variability in wood density drive divergences in Brazil's far north. iForest. 2023;**16**:95. DOI: 10.3832/ifor4137-016

[13] Camarero JJ, Hevia A. Links between climate, drought and minimum wood density in conifers. IAWA Journal. 2020;**41**:236-255. DOI: 10.1163/22941932-bja10005

[14] Szaban J, Jelonek T, Okí Nczyc A, Kowalkowski W. Results of a 57-year-long research on variability of wood density of the scots pine (*Pinus sylvestris* L.)

from different provenances in Poland. Forests. 2023;**14**:480. DOI: 10.3390/f14030480

[15] Hevia A, Campelo F, Chambel R, Vieira J, Alía R, Majada J, et al. Which matters more for wood traits in *Pinus halepensis* mill., provenance or climate? Annals of Forest Science. 2020;**77**:55. DOI: 10.1007/s13595-020-00956-y

[16] Burkhart HE, Amateis RL. Effects of early pruning on ring specific gravity in young loblolly pine trees. Wood and Fiber Science. 2020;**24**(52):139-151. DOI: 10.22382/wfs-2020-013

[17] Ruano A, Ruiz-Peinado R, Fernández-Golfín J, Hermoso E. Height growth for assessing core–outerwood transition on *Pinus sylvestris* and *Pinus nigra* Spanish stands. European Journal of Forest Research. 2020;**139**:273-278. DOI: 10.1007/s10342-019-01231-0

[18] Ruano A, Zitek A, Hinterstoisser B, Hermoso E. NIR hyperspectral imaging (NIR-HI) and μxRD for determination of the transition between juvenile and mature wood of *Pinus sylvestris* L. Holzforschung. 2019;**73**:621-627. DOI: 10.1515/hf-2018-0186

[19] Fries A, Ericsson T. Genetic parameters for earlywood and latewood densities and development with increasing age in scots pine. Annals of Forest Science. 2009;**66**:404-404. DOI: 10.1051/forest/2009019

[20] Mäkinen H, Hynynen J. Wood density and tracheid properties of scots pine: Responses to repeated fertilization and timing of the first commercial thinning. Forestry. 2014;**87**:437-447. DOI: 10.1093/forestry/cpu004

[21] Guller B, Isik K, Cetinay S. Variations in the radial growth and wood density components in relation to cambial age in 30-year-old *Pinus brutia* ten. at two test sites. Trees. 2012;**26**:975-986. DOI: 10.1007/s00468-011-0675-2

[22] del Río Gaztelurrutia M, Bravo Oviedo JA, Pretzsch H, Löf M, Ruiz-Peinado R. A review of thinning effects on scots pine stands: From growth and yield to new challenges under global change. Forest Systems. 2017;**26**:9. DOI: 10.5424/fs/2017262-11325

[23] Pretzsch H. Density and growth of forest stands revisited. Effect of the temporal scale of observation, site quality, and thinning. Forest Ecology and Management. 2020;**460**:117879. DOI: 10.1016/j.foreco.2020.117879

[24] Hevia A, Crabiffosse A, Álvarez-González JG, Ruiz-González AD, Majada J. Assessing the effect of pruning and thinning on crown fire hazard in young Atlantic maritime pine forests. Journal of Environmental Management. 2018;**205**:9-17. DOI: 10.1016/j.jenvman.2017.09.051

[25] Donovan VM, Roberts CP, Fogarty DT, Wedin DA, Twidwell D. Targeted grazing and mechanical thinning enhance forest stand resilience under a narrow range of wildfire scenarios. Ecosphere. 2022;**13**:e4061. DOI: 10.1002/ecs2.4061

[26] Reventlow DOJ, Nord-Larsen T, Skovsgaard JP. Pre-commercial thinning in naturally regenerated stands of European beech (*Fagus sylvatica* L.): Effects of thinning pattern, stand density and pruning on tree growth and stem quality. Forestry. 2019;**92**:120-132. DOI: 10.1093/forestry/cpy039

[27] York RA. Long-term taper and growth reductions following pruning intensity treatments in giant sequoia (*Sequoiadendron giganteum*). Canadian Journal of Forest Research.

2019;**49**:1189-1197. DOI: 10.1139/cjfr-2019-0118

[28] Amateis RL, Burkhart HE. Growth of young loblolly pine trees following pruning. Forest Ecology and Management. 2011;**262**:2338-2343. DOI: 10.1016/j.foreco.2011.08.029

[29] Mäkinen H, Hynynen J, Penttilä T. Effect of thinning on wood density and tracheid properties of scots pine on drained peatland stands. Forestry. 2015;**262**:359-367. DOI: 10.1093/forestry/cpv006

[30] Todaro L, Macchioni N. Wood properties of young Douglas-fir in Southern Italy: Results over a 12-year post-thinning period. European Journal of Forest Research. 2010;**130**:251-261. DOI: 10.1007/s10342-010-0425-9

[31] Carson SD, Cown D, McKinley R, Moore J. Effects of site, silviculture and seedlot on wood density and estimated wood stiffness in radiata pine at mid-rotation. New Zealand Journal of Forest Science. 2014;**44**:26. DOI: 10.1186/s40490-014-0026-3

[32] Gartner BL, Robbins JM, Newton M. Effects of pruning on wood density and tracheid length in young Douglas-fir. Wood and Fiber Science. 2005;**37**:304-313

[33] Sæbø JS, Socolar JB, Sánchez EP, Woodcock P, Bousfield CG, Uribe CAM, et al. Ignoring variation in wood density drives substantial bias in biomass estimates across spatial scales. Environmental Research Letters. 2022;**17**:054002. DOI: 10.1088/1748-9326/ac62ae

[34] Alfaro-Sánchez R, Jump AS, Pino J, Díez-Nogales O, Espelta JM. Land use legacies drive higher growth, lower wood density and enhanced climatic sensitivity in recently established forests. Agricultural and Forest Meteorology. 2019;**276-277**:107630. DOI: 10.1016/j.agrformet.2019.107630

[35] Moreno-Fernández D, Hevia A, Majada J, Cañellas I. Do common silvicultural treatments affect wood density of Mediterranean montane pines? Forests. 2018;**9**:80. DOI: 10.3390/f9020080

[36] Schweingruber FH. Tree Rings and Environment Dendroecology. Bern: Paul Haupt; 1996. p. 609

[37] Holmes RL. Computer-assisted quality control in tree ring dating and measurements. Tree-Ring Bulletin. 1983;**43**:69-78

[38] Zuur A, Ieno EN, Walker N, Saveliev AA, Smith GM. Mixed Effects Models and Extensions in Ecology with R. New York: Springer; 2009. p. 574

[39] Littell R, Pendergast J, Natarajan R. Tutorial in biostatistics: Modelling covariance structure in the analysis of repeated measures data. Statistics in Medicine. 2000;**19**:1793-1819

[40] Zuur A, Ieno E, Smith G. Analyzing Ecological Data. New York: Springer; 2007

[41] Wykoff WR. A basal area increment model for individual conifers in the Northern Rocky Mountains. Forest Science. 1990;**36**:1077-1104

[42] Vanclay JK. Mortality functions for North Queensland rain forests. Journal of Tropical Forest Science. 1991;**4**:15-36

[43] Mäkinen H, Isomäki A. Thinning intensity and growth of Norway spruce stands in Finland. Forestry. 2004;**77**:349-364. DOI: 10.1093/forestry/77.4.349

[44] Jaakkola T, Mäkinen H, Saranpää P. Wood density of Norway spruce: Responses to timing and intensity of first commercial thinning and fertilisation. Forest Ecology and Management. 2006;**237**:513-521. DOI: 10.1016/j.foreco.2006.09.083

[45] Pinheiro J, Bates D, Deb Roy S, Sarkar D, R Core Team. Nlme: Linear and Nonlinear Mixed Effects Models. R Package Version 4.31. 2023.

[46] Moreno-Fernández D, Cañellas I, Calama R, Gordo J, Sánchez-González M. Thinning increases cone production of stone pine (*Pinus pinea* L.) stands in the Northern Plateau (Spain). Annals of Forest Science. 2013;**70**:761-768. DOI: 10.1007/s13595-013-0319-3

[47] Mäkinen H, Isomäki A. Thinning intensity and growth of scots pine stands in Finland. Forest Ecology and Management. 2004;**201**:311-325. DOI: 10.1016/j.foreco.2004.07.016

[48] Moreno-Fernández D, Aldea J, Gea-Izquierdo G, Isabel I, Darío-Benito M. Influence of climate and thinning on *Quercus pyrenaica* Willd. Coppices growth dynamics. European Journal of Forest Research. 2021;**140**:187-197. DOI: 10.1007/s10342-020-01322-3

[49] Lloret F, Keeling EG, Sala A. Components of tree resilience: Effects of successive low-growth episodes in old ponderosa pine forests. Oikos. 2011;**120**:1909-1920. DOI: 10.1111/j.1600-0706.2011.19372.x

[50] Aldea J, Bravo F, Bravo-Oviedo A, Ruiz-Peinado R, Rodríguez F, del Río M. Thinning enhances the species-specific radial increment response to drought in Mediterranean pine-oak stands. Agricultural and Forest Meteorology. 2017;**237**:371-383. DOI: 10.1016/j.agrformet.2017.02.009

[51] Serrada R, Montero G, Reque JA. Compendio de Selvicultura Aplicada en España. Vol. 2008. Instituto Nacional de Investigación y Tecnología Agraria y Alimentaria. Ministerio de Educación y Ciencia. Madrid: Fundación Conde del Valle Salazar; 2008

[52] Bouriaud O, Leban JM, Bert D, Deleuze C. Intra-annual variations in climate influence growth and wood density of Norway spruce. Tree Physiology. 2005;**25**:651-660. DOI: 10.1093/treephys/25.6.651

[53] Nabais C, Hansen JK, David-Schwartz R, Klisz M, López R, Rozenberg P. The effect of climate on wood density: What provenance trials tell us? Forest Ecology and Management. 2018;**408**:148-156. DOI: 10.1016/j.foreco.2017.10.040

[54] Pritzkow C, Heinrich I, Grudd H, Helle G. Relationship between wood anatomy, tree-ring widths and wood density of *Pinus sylvestris* L. and climate at high latitudes in northern Sweden. Dendrochronologia. 2014;**32**:295-302. DOI: 10.1016/j.dendro.2014.07.003

Section 2

Species Traits and Wood Uses

Chapter 5

Quantifying: Genetic Traits in *Pinus wallichiana* Seedlings in the Northwestern Himalayan

Amanpreet Kaur and Rajesh Monga

Abstract

Pinus wallichiana, commonly known as the Himalayan blue pine, holds significant ecological and economic importance in the northwestern Himalayan region. Understanding the genetic traits and variability within its seedling population is essential for sustainable forest management and conservation efforts. This study aimed to quantify and assess the genetic traits of *Pinus wallichiana* seedlings within a nursery environment situated in the northwestern Himalayas. Our research involved the collection and analysis of data from a representative sample of *Pinus wallichiana* seedlings from different sites in Himachal Pradesh in 2019–2020. Results revealed a diverse genetic pool with notable heritability for key traits, highlighting the potential for selective breeding and genetic improvement programs. Furthermore, our findings provide valuable insights into the adaptation and resilience of *Pinus wallichiana* to changing environmental conditions, which is crucial for addressing the challenges posed by climate change. The quantification of genetic traits in this study not only enhances our understanding of the species but also offers practical applications for forest managers and policymakers in the region. This research contributes to the broader context of forest genetics and underscores the importance of genetic conservation efforts for the sustainable management of *Pinus wallichiana* in the northwestern Himalayas.

Keywords: Kail, genetic diversity, heritability, nursery, germination, and growth traits

1. Introduction

The Himalayan coniferous forests hold great importance due to their contributions in terms of timber resources, non-wood forest products, grazing areas, and the provision of habitats for endangered species. The Western Himalayan subalpine conifer forests, which cover an approximate area of 39,700 km^2 across India, Nepal, Pakistan, and Afghanistan, are primarily characterized by the presence of four dominant tree species. These species include *Pinus wallichiana* (blue pine), *Pinus gerardiana* (Chilgoza pine), *Abies pindrow* (fir), and *Picea smithiana* (spruce) [1]. The blue pine, a large evergreen tree, is widely distributed throughout the Himalayan region. In the country of Bhutan, the growth of this particular entity occurs at an elevation of

3400 m.a.s.l. [2, 3]. Seeds play a crucial role in the perpetuation of a species, although the likelihood of seed germination is frequently unknown and challenging to predict. The natural regeneration of many coniferous tree species is mostly dependent on seed-based mechanisms. Seeds, however, exhibit considerable variations, potentially attributable to variations in altitudinal ranges, necessitating the collection of seeds from different elevations [4]. The process of evaluating genetic parameters for qualities in *Pinus wallichiana* seedlings within a nursery entails examining several genetic and environmental factors that contribute to the observed variations in traits.

Genetic parameters offer valuable insights into several aspects of genetic features, including their heritability, genetic connection, and potential for selective breeding. In the field of quantitative genetics, the idea of heritability is crucial because it enables us to determine how much selection influences traits and how likely it is that breeding will pass those traits down to subsequent generations. The heritability of qualities determines the degree to which they are transmitted through successive generations [5]. However, it is important to note that solely relying on heritability estimates does not offer a comprehensive understanding of the degree to which development might be influenced by selection. Estimates of heritability alone do not show how much growth can be expected from selection. High genetic advance along with high heritability offers the most effective condition for selection of specific traits [6]. To what extent a trait improves in response to a certain selection pressure can be thought of as a genetic advance [7]. Genotype-phenotype analyses are indispensable for estimating the strength of associations between traits. Researchers can gain a better understanding of the relationships between various traits and how selection may affect them by studying genotype and phenotype correlations [8]. This knowledge allows for a more comprehensive understanding of the factors that contribute to growth and the potential impact of breeding on future generations. Additionally, examining these correlations can help identify any potential limitations or constraints in breeding programs that may affect the transmission of desired traits. The objectives of the present study encompass a comprehensive investigation into several key aspects: Our objective is to conduct a rigorous comparative analysis of the growth parameters exhibited by *Pinus wallichiana* seedlings collected from diverse altitudinal zones. This analysis will be carried out under real-world field conditions following a thorough nursery screening process by examining factors such as height, diameter, and foliage characteristics and then estimating the genetic parameters of *Pinus wallichiana*; we seek to quantify and assess the genetic diversity within the studied population. This aspect of the study is pivotal for understanding the species' genetic makeup, heritability of important traits, and the potential for genetic improvement and conservation efforts.

By pursuing these objectives, our research aims to provide a comprehensive understanding of how altitude influences the growth patterns of *Pinus wallichiana* seedlings while also shedding light on the genetic underpinnings of this ecologically and economically significant tree species.

2. Material and methods

Experimental site: The experiment was carried out during the years 2019–2020 at the forest nursery, Department of Silviculture and Agroforestry, Dr. Y.S. Parmar University of Horticulture and Forestry, Nauni, Solan, Himachal Pradesh, India. In 2019–2020, a study in Himachal Pradesh to gather seeds from various altitudinal ranges 1800–2100 (A_1), 2100–2400 (A_2), 2400–2700 (A_3), and > 2700 (A_4) m.a.s.l. [9].

The nursery location is situated at a height of 1200 m.a.s.l. in the northwestern region of the Himalayas, between the latitudes of 30°51′ N and longitudes of 76°11′ E. The experimental site exhibits topographical variations, including elevations, depressions, and a gradual incline in the southeastern direction. The region has a diverse range of temperatures, spanning from a minimum of 1°C during the winter season to a maximum of 33°C in the months of May and June, which are characterized as the peak of summer. The yearly precipitation ranges from 0 to 342 mm, with the highest amount occurring during the monsoon season, which typically spans from July to September.

Sample collection: Cones were collected from five phenotypically superior trees located in different sites of Himachal Pradesh (**Figure 1**). These trees displayed exceptional esthetic qualities and were selected based on a minimum distance of 100 m between each other. The mature cones of *Pinus wallichiana* were obtained from phenotypical superior trees in the intermediate stage of their life cycle and exhibited overall good health.

Prior to being sown, the seeds underwent a 60-day period of storage in a trench that consisted of discrete layers of sand and moss, which facilitated the process of breaking dormancy. Then, the 75 seeds per replication were subsequently seeded directly into root trainers that were filled with deodar forest soil mixed with FYM (Farmyard manure) at a ratio of 2:1. These root trainers were placed in polyhouse for 6 months and then in open nursery for hardening of the seedlings.

Design followed: Randomized Block Design.

Number of Replication: Three.

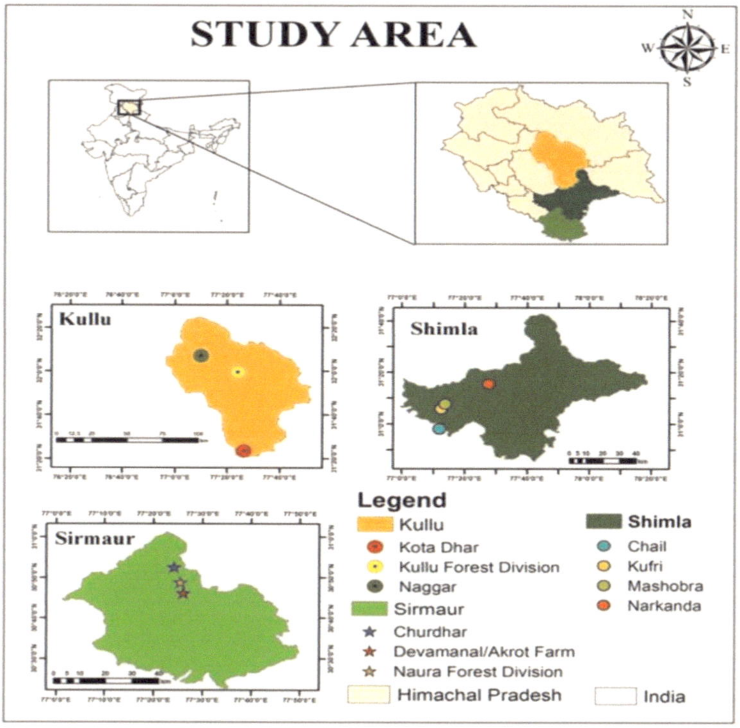

Figure 1.
Study area map.

Parameters: The observations (**Table 1**) were documented in seedlings that were 12 months old (final week of December). For calculation of genetic parameter irrespective of altitude, 60 plants were selected 5 plants per treatment, and for altitudinal effect, 15 plants per altitude were selected for data calculation.

Selecting individuals with optimal phenotypic expression is essential in the process of improving a trait. The developmental traits are the result of a combination of genetic and environmental variables. The computation of variability estimates and genetic parameters was performed for a range of seedling characteristics. The statistical measures of coefficient of variation and analysis of variance were computed for the phenotypic, genotypic, and environmental variations (**Table 2**). The study also involved the calculation of heritability estimates (in the broad sense), genetic advance (at a selection intensity of 5%), and genetic gain (expressed as a percentage of the mean). GCV and PCV can be categorized into three groups based on their magnitudes: low (less than 10%), moderate (10–20%), and high (more than 20%).

Statistical analysis: The data obtained from the field experiment was subjected to analysis using the analysis of variance (ANOVA) approach using OPSTAT software (http://opstat.pythonanywhere.com/) and MS Excel, which is commonly employed for examining the effects of two factors at p-value (0.05). The analysis of variance (ANOVA) procedures were used to test for significant effect of treatments (**Table 3**).

Traits	Method used
Germination (%)	$GP = \frac{Number\ of\ healthy\ seeds\ germinated}{Number\ of\ seeds\ sown} X\ 100$
Germination Capacity (%)	The cumulative number of seeds that germinated during the 28 days of test period plus the number of viable seeds at the end of the test expressed in percentage.
Germination Energy (%)	$GE = \frac{Number\ of\ healthy\ seeds\ germinated\ upto\ the\ time\ of\ peak\ germination}{Number\ of\ seeds\ sown} X\ 100$
Germination Speed	$GV = \sum \frac{DGS}{N} X\ (GP\ x\ 10)$ Where DGS = Daily germination speed = Cumulative germination per cent/Number of test days. N = Frequency or number of DGS during the test GP = Germination per cent at the end of the test
Germination Value	$GS = \sum_{i=1}^{n} \frac{number\ of\ newly\ germinated\ seeds}{number\ of\ days\ since\ sowing}$
Seedling Height (cm)	It was measured with the help of a meter scale in centimeter from leading shoot tip to the collar region of the seedling at the ground surface.
Collar Diameter (mm)	The diameter of seedling was recorded in millimeters with the help of a digital Vernier caliper.
Needle Area (cm^2)	The projected leaf area was determined on CI-202 Leaf Area Meter.
Root Length (cm)	The root length was measured with the help of using a digital caliper (Mitutoyo Absolute) from the cut base to the tip of the taproot.
Seedling Dry Weight (gm)	Total seedling dry weight was obtained by summing up root and shoot dry weight in gram.
Survival (%)	$SP = \frac{\text{Total percent germination}}{\text{Number of seeds germinated in the that seed source}} \times 100$

Table 1.
Germination and growth parameters measured in the nursery.

Sr. No.	Parameter name	Symbol	Calculation	Description
1.	Genotypic variance	Vg	$\frac{Mt-Me}{R} \times 100$	Mt = Mean sum of square due to treatment Me = Mean sum of square due to error R = Number of replications. X = Mean of the character K = Selection intensity at 5 percent, which is equal to 2.06
2.	Phenotypic variance	Vp	Vg + Ve	
3.	Environmental variance	Ve	Me	
4.	Phenotypic coefficient of variation	PCV (%)	$\frac{\sqrt{Vp}}{X} \times 100$	
5.	Genotypic coefficient of variance	GCV (%)	$\frac{\sqrt{Vg}}{X} \times 100$	
6.	Environmental coefficient of variance	ECV (%)	$\frac{\sqrt{Ve}}{X} \times 100$	
7.	Heritability in broad sense	h^2 (%)	$\frac{Vg}{Vp}$	
8.	Genetic advance	GA	K. $\sqrt{VP^*} h^2$	
9.	Genetic gain/Genetic advance as percent of mean	GG	$\frac{GA}{X} \times 100$	

Table 2.
Genetic parameters calculation methods.

Source of variation	Degree of freedom	Sum of squares	Mean sum of squares	F. Cal.
Replications	(r-1)	S_r	$\frac{Sr}{(r-1)} = Mr$	M_r/M_e
Genotype	(g-1)	S_g	$\frac{St}{(g-1)} = Mt$	M_g/M_e
Error	(r-1) (g-1)	S_{rg}	$\frac{Se}{(r-1)(g-1)} = Me$	
Total	(rg-1)	$S_{(rg-1)}$		

Where r = Number of replications, g = Number of genotypes/Plant number, and M = Mean sum of square.
The calculated "F" values were compared with the tabulated "F" values at 5 percent level of significance. If the calculated value is higher than the tabulated value, it will be considered significant. Critical difference (CD) for comparing the means of any two treatments will be calculated as: SE (d)= ± (2Me/r) ½
*Critical Difference (CD) = SE (d) × t (5%) value at error degrees of freedom. The calculation of predicted genetic advance at a selection intensity of 5 percent was performed using the formulas discussed in **Table 2** [10], taking into account genotypic and phenotypic variances, environmental variances, and coefficients of variability.*

Table 3.
Statistical analysis of the measured germination and growth traits.

3. Results

3.1 Assessment of genotypic and phenotypic variability in nursery seedlings

The study aimed to assess the variability in germination and seedling growth characteristics among different seed sources, taking into account both genetic and environmental factors. A sample of five mother trees was used for cone and seed collection, and the results are presented in **Table 4**.

The data in **Table 4** indicate significant variation (p-value $\leq$0.05) in various characteristics among the seed sources. Notably, the survival rate, needle area, root length, and germination rate exhibited the highest levels of environmental diversity. This suggests that these traits are particularly influenced by environmental conditions.

Traits	Variance			Coefficient of variance			HBS	GA	GAPM	CD (0.05)
	EV	GV	PV	ECV	PCV	GCV				
GC	6.29	75.6	81.9	4.04	14.6	14.0	0.92	17.1	27.6	1.96
GP	7.56	70.1	77.6	4.61	14.8	14.0	0.90	16.3	27.3	2.15
GE	2.80	4.37	7.18	4.88	7.81	6.10	0.61	3.35	9.76	1.31
GV	0.20	3.41	3.61	3.20	13.7	13.3	0.95	3.68	26.5	0.35
GS	0.10	0.74	0.84	10.3	29.1	27.2	0.88	1.65	52.3	0.25
SH	0.73	3.31	4.04	8.72	20.5	18.5	0.82	3.37	34.4	0.67
CD	0.06	1.16	1.21	8.13	37.9	37.0	0.95	2.15	74.2	0.18
NA	12.4	25.7	38.1	10.2	17.9	14.7	0.67	8.52	24.8	2.15
RL	7.59	16.8	24.4	17.4	31.2	25.9	0.69	6.97	44.1	2.75
SDW	0.01	0.04	0.05	24.3	49.1	42.6	0.75	0.34	75.9	0.09
SP	28.9	62.5	91.4	11.3	20.2	16.7	0.68	13.4	28.3	4.20

p-value (0.05) = <0.0001.
EV, Environment variance; PV, Phenotypic variance; GV, Genotypic variance; ECV, Environment coefficient of variance; PCV, Phenotypic coefficient of variance; GCV, Genotypic coefficient of variance, HBS, Broad-sense heritability; GA, Genetic advance; GAPM, Genetic advance as percent of mean; GC, Germination capacity; GP, Germination percent; GC, Germination capacity; GE, Germination energy; GS, Germination speed; GV, Germination value; SH, Seedling height; CD, Collar Diameter; NA, Needle area; RL, Root length; SDW, Seedling Dry weight; SP, Survival percentage.

Table 4.
Genetic parameters analysis of the Pinus wallichiana seedlings, irrespective of altitudinal ranges.

The study also revealed substantial variability in both genotypic and phenotypic characteristics, with germination capacity and plant survival showing the most significant variation. Additionally, the coefficient of variation for environmental, phenotypic, and genotypic components showed higher values for the dry weight of seedlings, indicating that this trait is influenced by a combination of genetic and environmental factors. On the other hand, collar diameter and germination value displayed the highest levels of broad-sense heritability, indicating a strong genetic component in these traits. Genetic advance and genetic gain were most prominent in germination capacity and germination percentage, whereas dry weight of seedlings and collar diameter showed the greatest potential for genetic improvement.

3.2 Assessment of genotypic and phenotypic variability across altitudinal gradients in nursery seedlings

Table 5 presents data illustrating significant variations (p-value ≤0.05) in germination and growth parameters across different altitudinal ranges. Environmental variance was notably high in survival percentage at altitude A_4, highlighting the influence of altitude on this particular trait. Altitude A_2 exhibited the highest genotypic and phenotypic variations in germination percentage, indicating the significance of altitude in shaping these characteristics. Furthermore, the coefficients of variance for environmental, genotypic, and phenotypic components were highest for the dry

	ALT	GC	GP	GE	GV	GS	SH	CD	NA	RL	SDW	SP
EV	A_1	5.12	3.28	0.59	0.07	0.05	0.87	0.04	0.47	13.9	0.01	14.4
	A_2	5.48	4.06	0.80	0.22	0.12	0.35	0.07	12.9	4.87	0.01	13.2
	A_3	4.04	3.48	3.27	0.14	0.07	0.39	0.07	9.83	3.20	0.02	31.4
	A_4	2.49	0.85	2.46	0.32	0.08	0.35	0.01	1.97	2.95	0.01	34.5
PV	A_1	14.1	17.5	1.20	0.35	0.17	0.87	0.15	43.7	22.3	0.05	20.8
	A_2	79.1	81.3	8.18	4.23	0.69	4.86	1.03	43.7	12.6	0.04	14.8
	A_3	22.5	15.1	4.98	1.07	0.99	1.25	0.73	17.8	4.17	0.08	58.0
	A_4	13.1	6.96	5.89	7.57	1.45	3.00	2.85	13.8	9.18	0.03	58.9
GV	A_1	8.94	14.2	0.61	0.27	0.12	NA	0.11	43.2	8.48	0.04	6.42
	A_2	73.6	77.3	7.38	4.01	0.57	4.50	0.96	30.8	7.75	0.03	1.60
	A_3	18.4	11.6	1.71	0.93	0.92	0.86	0.67	7.95	0.97	0.06	26.6
	A_4	10.6	6.11	3.43	7.25	1.37	2.64	2.83	11.8	6.23	0.02	24.4
ECV	A_1	3.12	2.61	2.20	1.87	6.37	7.92	7.56	1.82	17.6	18.7	6.56
	A_2	3.66	3.26	2.48	3.30	10.9	6.67	10.1	10.1	13.8	23.8	7.27
	A_3	3.51	3.38	5.45	2.60	8.84	6.34	8.15	9.97	12.8	27.0	13.4
	A_4	2.90	1.77	4.75	4.46	9.50	6.81	3.52	4.23	14.3	25.7	14.7
PCV	A_1	5.17	6.01	3.14	4.06	12.1	7.90	14.5	17.6	22.3	52.0	7.89
	A_2	13.9	14.6	7.92	14.61	25.7	24.7	38.9	18.6	22.2	42.4	7.70
	A_3	8.28	7.05	6.73	7.28	33.2	11.3	27.2	13.4	14.6	56.9	18.2
	A_4	6.66	5.06	7.35	21.57	41.0	19.9	52.5	11.2	25.2	40.3	19.2
GCV	A_1	4.12	5.42	2.23	3.61	10.2	NA	12.3	17.5	13.8	48.5	4.38
	A_2	13.4	14.2	7.52	14.2	23.3	23.8	37.6	15.6	17.4	35.1	2.53
	A_3	7.50	6.18	3.94	6.79	32.0	9.38	25.9	8.96	7.02	50.1	12.3
	A_4	5.99	4.74	5.61	21.10	39.8	18.6	52.4	10.4	20.8	31.1	12.3

	ALT	GC	GP	GE	GV	GS	SH	CD	NA	RL	SDW	SP
HBS	A_1	0.64	0.81	0.51	0.79	0.72	NA	0.73	0.99	0.38	0.87	0.31
	A_2	0.93	0.95	0.90	0.95	0.82	0.93	0.93	0.71	0.61	0.69	0.11
	A_3	0.82	0.77	0.34	0.87	0.93	0.69	0.91	0.45	0.23	0.77	0.46
	A_4	0.81	0.88	0.58	0.96	0.95	0.88	1.00	0.86	0.68	0.59	0.41
GA	A_1	4.91	6.99	1.14	0.96	0.62	NA	0.57	13.5	3.69	0.40	2.90
	A_2	17.1	17.7	5.31	4.02	1.40	4.21	1.95	9.60	4.50	0.28	0.86
	A_3	8.01	6.16	1.58	1.86	1.91	1.58	1.60	3.88	0.98	0.46	7.20
	A_4	6.03	4.77	2.91	5.43	2.34	3.15	3.46	6.56	4.24	0.20	6.56
GAPM	A_1	6.77	10.1	3.28	6.59	17.9	NA	21.7	35.8	17.4	93.2	5.02
	A_2	26.6	28.6	14.7	28.6	43.5	47.3	74.7	27.1	28.1	59.9	1.72
	A_3	14.0	11.18	4.76	13.1	63.5	16.0	50.9	12.3	6.97	90.8	17.2
	A_4	11.1	9.16	8.82	42.5	79.8	36.1	107	19.8	35.2	49.3	16.4
p-value (0.05)	A_1	<0.0001	<0.0001	0.0008	<0.0001	<0.0001	0.4957	<0.0001	<0.0001	0.0092	<0.0001	0.0269
	A_2	<0.0001	<0.0001	<0.0001	<0.0001	<0.0001	<0.0001	<0.0001	<0.0001	<0.0001	<0.0001	0.2344
	A_3	<0.0001	<0.0001	0.0163	<0.0001	<0.0001	<0.0001	<0.0001	0.0027	0.0709	0.05	0.0021
	A_4	<0.0001	<0.0001	0.0001	<0.0001	<0.0001	<0.0001	<0.0001	<0.0001	<0.0001	<0.0001	0.005
CD (0.05)	A_1	3.14	2.52	1.07	0.38	0.31	1.30	0.28	5.18	0.95	0.11	5.27
	A_2	3.25	2.80	1.25	0.65	0.49	0.83	0.37	3.07	4.98	0.15	5.05
	A_3	1.61	1.50	1.45	0.30	0.21	0.50	0.21	1.44	2.52	0.11	4.49
	A_4	2.19	1.28	2.18	0.79	0.39	0.82	0.16	2.39	1.95	0.14	8.16

EV, Environment variance; PV, Phenotypic variance; GV, Genotypic variance; ECV, Environment coefficient of variance; PCV, Phenotypic coefficient of variance; GCV, Genotypic coefficient of variance; HBS, Broad-sense heritability; GA, Genetic advance; GAPM, Genetic advance as percent of mean; ALT, Altitude; GC, Germination capacity; GP, Germination percent; GC, Germination capacity; GE, Germination energy; GS, Germination speed; GV, Germination value; SH, Seedling height; CD, Collar diameter; NA, Needle area; RL, Root length; SDW, Seedling dry weight; SP, Survival percentage

Table 5.
Genetic parameter analysis of germination and growth traits of Pinus wallichiana along different altitudinal ranges.

weight of seedlings at altitude A_3, suggesting that this trait is influenced by both genetic and environmental factors, particularly at this altitude. Collar diameter at altitude A_4 displayed the greatest degree of heritability, emphasizing the strong genetic influence on this trait. Needle area at altitude A_1 also showed a high level of heredity. The study revealed that altitude A_2 had the most significant genetic enhancement in germination percentage compared to other altitudes, while the most substantial genetic improvement was observed in the dry weight of seedlings at altitude A_1. The disparity between phenotypic coefficient of variation (PCV) and genotypic coefficient of variation (GCV) across all traits underscores the interplay between genetic and environmental factors in shaping these characteristics. This study highlights the pronounced influence of environmental factors in the A_1 altitudinal range on seedling growth and development, emphasizing the importance of considering both genetic and environmental elements in the analysis of these traits.

4. Discussion

The findings of this study suggest that root length, survival percentage, and germination percentage were predominantly influenced by site-specific and environmental factors. While heredity plays a significant role, it is clear that site conditions exerted a greater influence on these traits. To gain a comprehensive understanding of the potential for selection in the studied material, it is essential to examine heritability and expected genetic gain together, as these factors collectively provide a more reliable and credible assessment [10]. Relying solely on heredity can be limiting, and incorporating both heritability estimates and genetic gain into predictions for the future selection of optimal genotypes offers a more advantageous approach [11].

Previous research on *Grewia optiva* [12] and *Celtis australis* [13] has documented higher heritability estimates and statistically significant genetic effects on seed weight increase. Furthermore, significant findings in *Albizia chinensis* [14], *Celtis australis* [13], and *Pinus wallichiana* [15] align with the outcome of our study, supporting the influence of genetics on these traits.

Our investigation revealed that genotypic coefficients of variance (GCV) were consistently higher than their corresponding phenotypic coefficients of variance (PCV), indicating that the genetic component played a more substantial role in trait variation at the genotypic level. This finding is in agreement with other studies, such as [16], who observed inter- and intraspecific genetic variation in seedling drought tolerance in different pine species. Similarly, [17] identified distinct patterns in loci associated with phenotypic traits in loblolly pine, suggesting a genetic basis for these traits. Genetic correlations tended to be higher than phenotypic correlations, indicating that phenotypic correlations can serve as fair estimates of their genetic counterparts in many scenarios [18]. Additionally, a negative genetic correlation between height growth and other traits in jack pine, emphasizing the importance of considering multiple traits in selection approaches [1]. GCV is a superior indicator of the genetic relationships among traits in pines [19].

This phenomenon may be attributed to genotype–environment interaction or the influence of environmental elements on trait expression. Our findings align with similar observations in full-sibling offspring of specifically chosen clones of *Populus deltoides* [20] and the outcomes of studies on willow clones [21], collectively suggesting that GCV provides a more accurate representation of genetic relationships among traits than PCV. The results of studies conducted by various scientists [11, 22, 23] are consistent with these findings. The observation of a notable level of agreement between phenotypic and genotypic coefficients of variation suggests substantial diversity in genotypes, indicating the potential for improvement in these specific characteristics. This underscores the importance of considering genotypic values when studying trait inheritance and selection in plant populations.

5. Conclusion

In conclusion, our research uncovers substantial genetic diversity within *Pinus wallichiana* genotypes. While environmental factors predominantly dictate survival rates, our study underscores the influential role of genetic and phenotypic variance, particularly in shaping germination capacity and, notably, germination percentage. Among these traits, germination percentage emerges as the most genetically diverse, suggesting its potential as a prime target for selective breeding initiatives. Furthermore, our findings highlight altitudes A_1 and A_2 as regions where heritability plays a prominent role in governing germination and growth traits, indicating their relative resilience to environmental influences within these altitudinal ranges. Consequently, focusing on trait improvement within the lower (1800–2100 m.a.s.l.) and mid-altitudinal (2100–2400 m.a.s.l.) zones appears strategically advantageous. The application of these insights in selective breeding holds promise for bolstering the adaptability and vitality of *Pinus wallichiana* populations. As we look ahead, our findings offer valuable guidance for sustainable forest management, conservation, and breeding programs, particularly in diverse altitudinal contexts, fostering the long-term resilience of this important species.

Acknowledgements

The authors express their gratitude for the help provided by Dr. YS Parmar, the Head of the Department of Silviculture and Agroforestry, University of Horticulture and Forestry, in relation to the current study.

Conflict of interest

A researcher has a significant financial interest (publication fee).

A. Appendix

See **Table 6**.
See **Table 7**.
See **Table 8**.

		Germination capacity			Germination percent			Germination energy			Germination value			Germination speed		
SOV	**Df**	**SS**	**MSS**	**F cal**	**SS**	**MSS**	**F cal**	**SS**	**MSS**	**F cal**	**SS**	**MSS**	**F cal**	**SS**	**MSS**	**F cal**
Genotype	59.00	13750.1	233.05	37.0	12849.0	217.78	28.81	939.70	15.93	5.68	616.00	10.44	52.91	136.70	2.32	22.19
Replication	2.00	167.98	83.99	13.34	742.84	371.42	49.14	932.70	466.35	166.37	14.52	7.26	36.80	0.35	0.17	1.65
Error	118.00	742.73	6.29		891.85	7.56		330.77	2.80		23.29	0.20		12.32	0.10	
Total	179.0	14660.8			14483.7			2203.2			653.82			149.37		
		Seedling height			Collar diameter			Root length			Needle area			Seedling dry weight		
SOV	Df	SS	MSS	F cal	SS	MSS	F cal	SS	MSS	F cal	SS	MSS	F cal	SS	MSS	F cal
Genotype	59.00	7.15	0.12	10.21	208.30	3.53	63.28	3416.31	57.90	7.63	5274.71	89.40	7.20	7.15	0.12	10.21
Replication	2.00	0.04	0.02	1.68	0.72	0.36	6.43	755.97	377.98	49.80	78.41	39.20	3.16	0.04	0.02	1.68
Error	118.00	1.40	0.01		6.58	0.06		895.61	7.59		1465.84	12.42		1.40	0.01	
Total	179.0	8.59			215.60			5067.9			6818.9			8.59		
		Survival Percentage														
SOV	Df	SS	MSS	F cal												
Genotype	59.00	12766.6	216.38	7.49												
Replication	2.00	1101.90	550.95	19.07												
Error	118.0	3409.73	28.90													
Total	179.0	17278.2														

Table 6.
ANOVA for nursery traits in Pinus wallichiana *seedling.*

Altitudes	SOV	Df	Germination capacity			Germination percent			Germination energy			Germination value			Germination speed		
			SS	MSS	F cal	SS	MSS	F cal	SS	MSS	F cal	SS	MSS	F cal	SS	MSS	F cal
1800-2100 masl (A1)	Genotype	14	447.07	31.933	6.24	641.97	45.855	13.963	33.83	2.416	4.082	12.566	0.898	12.107	5.886	0.420	8.745
	Replication	2	100.005	50.00	9.771	578.7	289.03	88.01	163.84	81.918	138.390	8.664	4.332	58.434	0.262	0.131	2.724
	Error	28	143.293	5.118		91.956	3.284		16.574	0.592		2.076	0.074		1.346	0.048	
	Total		690.363			1311.9			214.23			23.306			7.493		
			Germination capacity														
2100-2400 masl (A2)	Genotype	14	3168.167	226.29	41.328	3302.642	235.90	58.155	321.148	22.939	28.567	171.455	12.247	56.812	25.487	1.820	14.783
	Replication	2	43.318	21.659	3.956	210.356	105.178	25.929	613.53	306.76	382.030	1.626	0.813	3.772	0.039	0.019	0.157
	Error	28	153.319	5.476		113.580	4.056		22.484	0.803		6.036	0.216		3.448	0.123	
	Total		3364.80			3626.578			957.162			179.117			28.973		
			Germination capacity														
2400-2700 masl (A3)	Genotype	14	830.703	59.336	14.676	536.968	38.355	11.029	117.691	8.407	2.569	41.060	2.933	21.419	39.764	2.840	40.207
	Replication	2	168.076	84.038	20.785	230.559	115.279	33.148	100.70	50.350	15.384	2.659	1.330	9.711	0.241	0.121	1.709
	Error	28	113.209	4.043		97.376	3.478		91.641	3.273		3.834	0.137		1.978	0.071	
	Total		1111.988			864.902			310.031			47.553			41.984		
			Germination capacity %														
>2700masl (A4)	Genotype	14	479.44	34.245	13.761	268.412	19.172	22.572	178.622	12.759	5.186	308.98	22.070	68.139	58.514	4.180	53.766
	Replication	2	119.810	59.905	24.072	289.016	144.51	170.135	185.821	92.911	37.764	3.847	1.923	5.938	3.177	1.588	20.431
	Error	28	69.681	2.489		23.782	0.849		68.888	2.460		9.069	0.324		2.177	0.078	
	Total		668.93			581.210			433.331			321.899			63.867		

Table 7.
ANOVA for germination traits in Pinus wallichiana *Seedlings among different altitudinal ranges.*

Altitudes	SOV	Df	Root length			Seedling height			Collar diameter			Seedling dry weight			Needle area			Survival Percent		
			SS	MSS	F cal	SS	MSS	F cal	SS	MSS	F cal	SS	MSS	F cal	SS	MSS	F cal	SS	MSS	F cal
1800-2100m asl (A1)	Genotype	14	550.219	39.301	2.833	11.968	0.855	0.981	5.042	0.36	8.988	1.879	0.134	21.17	1821.78	130.13	277.78	470.796	33.628	2.342
	Replication	2	75.204	37.602	2.710	23.941	11.971	13.74	0.134	0.07	1.676	0.02	0.01	1.877	20.833	10.417	22.23	182.221	91.110	6.345
	Error	28	388.470	13.874		24.402	0.872		1.122	0.04		0.178	0.01		13.117	0.468		402.05	14.359	
	Total		1013.89			60.312			6.298			2.08			1855.73			1055.07		
2100-2400m asl (A2)	Genotype	14	393.850	28.132	5.777	194.09	13.86	39.32	41.49	2.96	42.294	1.283	0.09	7.539	1473.56	105.25	8.183	251.848	17.989	1.364
	Replication	2	523.209	261.60	53.72	54.299	27.15	76.99	0.589	0.29	4.201	0.01	0.01	0.142	172.800	86.40	6.717	264.662	132.33	10.03
	Error	28	136.347	4.870		9.873	0.353		1.962	0.07		0.34	0.01		360.150	12.863		369.289	13.189	
	Total		1053.41			258.27			44.04			1.627			2006.51			885.799		
2400-2700m asl (A3)	Genotype	14	85.570	6.112	1.908	41.432	2.959	7.551	28.917	2.07	31.364	2.921	0.21	11.29	471.486	33.678	3.426	1557.76	111.269	3.545
	Replication	2	75.881	37.941	11.84	15.569	7.784	19.86	1.242	0.62	9.428	0.08	0.04	2.248	323.408	161.70	16.448	599.384	299.69	9.549
	Error	28	89.696	3.203		10.974	0.392		1.844	0.07		0.517	0.02		275.267	9.831		878.787	31.385	
	Total		251.148			67.975			32.00			3.521			1070.16			3035.93		
>2700m asl (A4)	Genotype	14	302.950	21.639	7.331	115.94	8.281	23.52	119.16	8.512	664.86	0.79	0.06	5.375	524.156	37.440	18.968	1509.37	107.81	3.126
	Replication	2	280.123	140.06	47.45	53.846	26.92	76.5	0.049	0.02	1.933	0.00	0.00	0.02	323.408	161.70	81.925	849.636	424.82	12.32
	Error	28	82.647	2.952		9.857	0.352		0.358	0.01		0.29	0.011		55.267	1.974		965.595	34.486	
	Total		665.720			179.64			119.57			1.09			902.831			3324.60		

Table 8.
ANOVA for growth traits in Pinus wallichiana *Seedlings among different altitudinal range.*

Author details

Amanpreet Kaur[1]* and Rajesh Monga[2]

1 Silviculture and Forest Management Division, Forest Research Institute, Dehradun, Uttarakhand, India

2 Clean Development Mechanism, KBSCertification Services, Faridabad, Haryana, India

*Address all correspondence to: amanjambh59@gmail.com; dr.amanjambh@gmail.com

References

[1] Ahmed M, Hussain T, Sheikh AH, Siddiqui MF. Phytosociology and structure of Himalayan forests from different climatic zones of Pakistan. Pakistan Journal of Botany. 2006;**38**: 361-383

[2] Dar JA, Sundarapandian P. Patterns of plant diversity in seven temperate forest types of Western Himalaya, India. Journal of Asia–Pacific Biodiversity. 2016;**9**:280-292. DOI: 10.1016/j.japb.2016.03.018

[3] Farjon A. *Pinus wallichiana*. The IUCN red list of threatened species. 2013. DOI: 10.2305/iucn.uk.2013-1.rlts.t191650a1991477.en

[4] Barnett PE, Farmer RE. Altitudinal variation in germination characteristics of yellow poplar in the southern Appalachians. Silvae Genetica. 1978;**27**: 101-104

[5] Sabesan T, Suresh R, Saravanan K. Genetic variability and correlation for yield and grain quality characters of rice grown in coastal saline low land of Tamil Nadu. Electronic Journal of Plant Breeding. 2009;**1**:56-59

[6] Islam M, Khalequzzaman M, Bashar M, Ivy N, Haque MM, Mian MAK. Variability assessment of aromatic and fine rice germplasm in Bangladesh based on quantitative traits. The Scientific World Journal. 2016;**2016**:1-14. DOI: 10.1155/2016/2796720

[7] Ogunniyan DJ, Olakojo SA. Genetic variation, heritability, genetic advance and agronomic character association of yellow elite inbred lines of maize (*Zea mays* L.). Nigerian Journal of Genetics. 2014;**28**:24-28. DOI: 10.1016/j.nigjg.2015.06.005

[8] Zhao XY, Ma KF, Shen YB, Zhang M, Li KY, Wu R. Characteristic variation and selection of forepart hybrid clones of Sect. Populus. Journal of Beijing Forestry University. 2012;**34**:45-51

[9] Kaur AP, Monga R, Bhardwaj DR, Sharma JP. Estimation of Genetic Parameters of *Pinus wallichiana* Seedlings in the Nursery. International Journal of Bio-resource and Stress Management. 2022;**13**(6):578-585

[10] Johnson HW, Robinson HF, Comstock RF. Estimates of genetic and environmental variability in soyabean. Agronomy Journal. 1955;**47**:314-318. DOI: 10.2134/agronj1955.00021962004700070009xx

[11] Volker PW, Dean CA, Tibbits WN, Ravenwood IC. Genetic parameters and gain expected from selection in *Eucalyptus globulus* in Tasmania. Silvae Genetica. 1990;**39**:18-21

[12] Uniyal AK. Provenance variation in seed and seedling of Grevia optiva Drumm. [PhD thesis] H.N.B.G.U. Srinagar Garhwal, Uttarakhand, India; 1998

[13] Singh B, Bhatt BP, Prasad P. Effect of seed source and temperature on seed germination of *Celtis australis* (L.): A promising agroforestry tree crop of central Himalaya, India. Forests Trees and Livelihood. 2004;**14**:53-60. DOI: 10.1080/14728028.2004.97524799

[14] Dhanai CS. Provenance variation in seed and seedling of *Albizia chinensis* (Osbeck). Ph.D. Thesis work. H.N.B.G. U. Srinagar Garhwal, Uttarakhand; 2002

[15] Rawat K, Bakshi K. Provenance variation in cone, seed and seedling characteristics in natural populations of

Pinus wallichiana A.B. Jacks (Blue Pine) in India. Annals of Forest Research. 2011; **54**:39-55

[16] Andres GF. Phenotypic variation among natural populations of pines: Implications for the management and conservation of genetic resources [Doctoral thesis]. Department University Institute of Research in Sustainable Forest Management. 2018

[17] Eckert AJ, Wegrzyn JL, Liechty JD, Lee JM, Patrick Cumbie W, Davis JM, et al. The evolutionary genetics of the genes underlying phenotypic associations for loblolly pine (*Pinus taeda*, Pinaceae). Genetics. 2013;**195**(4):1353-1372. DOI: 10.1534/genetics.113.157198

[18] James M, Cheverud. A comparison of genetic and phenotypic correlations. Evolution. 1988;**42**(5):958-968. DOI: 10.1111/j.1558-5646.1988.tb02514.x

[19] Shalizi MN, Gezan SA, McKeand SE, Sherrill JR, Cumbie WP, Whetten RW, et al. Correspondence between breeding values of the same *Pinus taeda* L. genotypes from clonal trials and half-sib seedling progeny trials. Forest Science. 2020;**66**(5):600-611. DOI: 10.1093forsci/fxaa016

[20] Kadam SK. Evaluation of full-sib progenies of selected clones of poplar (*Populus deltoides* Bartr.) [PhD thesis]. Dehradun: Forest Research Institute; 2002

[21] Singh NB, Sharma JP, Huse SK, Thakur IK, Gupta RK, Sankhyan HP. Heritability, genetic gain, correlation and principal component analysis in introduced willow (Salix species) clones. Indian Forester. 2012;**138**:1100-1109

[22] Reni YP, Rao YK. Genetic variability in soybean (*Glycine max* (L) Merrill). International Journal of Plant, Animal and Environmental Sciences. 2013;**3**: 35-38

[23] Showkat M, Tyagi SD. Genetic variability in soybean (*Glycine max* L. Merrill). Research. Journal of Agricultural Science. 2010;**1**:102-106

Chapter 6

Ethnobotany of Conifers in the Philippines

Richard Clemente

Abstract

Gymnosperms are a few of the groups of plants that are often neglected. Primary and secondary literature have been consulted to establish listings of the recognized gymnosperms in the Philippines. About seven known gymnosperm families and thirteen genera of conifers are found in the Philippines. To wit, two genera from Araucariaceae, one genus from Cycadaceae, one from Gnetaceae, one from Pinaceae, six from Podocarpaceae, one from Phyllocladaceae, and one from Taxaceae. *Agathis dammara* (Lamb.) Poir. And *Agathis philippinensis* Warb. (Araucariaceae), *Pinus kesiya* Royle ex Gordon and *Pinus merkusii* Jungh. & de Vriese (Pinaceae), *Gnetum gnemon* L. and *Gnetum latifolium* Blume. (Gnetaceae), *Dacrycarpus imbricatus* (Blume) de Laub., *Podocarpus macrocarpus* de Laub. and *Sundacarpus amarus* (Blume) C. N. Page (Podocarpaceae) and *Phyllocladus hypophyllus* Hook.f. (Phyllocladaceae) are gymnosperms documented with ethnobotanical knowledge. Other species warrant further research on their economic value and must be explored. The conservation status of these conifers should be known to all.

Keywords: native, endemic, conservation status, gymnosperms, economic value

1. Introduction

It is widely agreed that gymnosperms can resolve many environmental problems, such as pollution, soil erosion, and desertification. In achieving the United Nations Sustainable Development Goals (SDGs), ethnobotanical knowledge can play a vital role in addressing global concerns such as poverty alleviation, food security, food safety and availability, sustainable consumption and production, mitigate climate crisis and its impacts, biodiversity conservation and establishing networks between indigenous peoples. Ethnobotanists have played in unraveling and documenting plant-people interactions and unlocking the knowledge by research. People has been utilizing plants since ancient times, primarily on food and medicinal purposes and yet few are still undocumented. Gymnosperms are among the plant groups often neglected in research in the Philippines. About seven known gymnosperm families and thirteen genera of conifers are found (**Table 1**). Most of its economic value has no information available and can be a big area of interest. This chapter aims to review the ethnobotany of conifers in the Philippines.

Families, Genera		Number of species
Family Araucariaceae	Genus *Agathis*	2
	Genus *Araucaria*	2
Family *Cycadaceae*	Genus *Cycas*	14
Family Gnetaceae	Genus *Gnetum*	4
Family Pinaceae	Genus *Pinus*	2
Family Podocarpaceae	Genus *Dacrycarpus*	2
	Genus *Dacrydium*	4
	Genus *Falcatifolium*	1
	Genus *Nageia*	1
	Genus *Podocarpus*	10
	Genus *Sundacarpus*	1
Family Phyllocladaceae	Genus *Phyllocladus*	1
Family Taxaceae	Genus *Taxus*	1
Total		**45**

[1]*Includes all the native and cultivated, not naturalized species.*

Table 1.
List of species of gymnosperms found in the Philippines.[1]

2. Materials and methods

Descriptive method was employed by consulting primary and secondary literature to established listings of the recognized gymnosperms in the Philippines. Online resources, *The Gymnosperm Database* [1] and *Co's Digital Flora of the Philippines* [2], an authority in listing all the vascular plants in the country served as the framework of this paper. Research articles, book chapters, current news and compilations provide additional information.

3. Results

A taxonomic checklist of conifers were presented based on their family names with description, distribution, conservation status and ethnobotanical information.

3.1 Family Araucariaceae Henkel & W. Hochstetter

Family Araucariaceae has two genera, namely, *Agathis* and *Araucaria*.

Genus *Agathis* Salisb.

Agathis is Greek word meaning 'a ball of thread', an allusion of its globose female cone. The word 'kauri' is of Maori origin, applied by that people to *Agathis australis*, a common species and generalized in term to all species of *Agathis* [3]. Throughout its distribution, *Agathis* is one of the highly sought after sources of straight-grained, easily worked timber [4]. Nowadays, it is relativity scarce but still of premium value, it has been largely logged out. Its current production is usually derived from plantations. Genus *Agathis* has two species, namely *Agathis dammara* and *A. philippinensis.*

Agathis dammara (Lamb.) Rich. & A.Rich. ex A.Rich. a native species is distributed in Palawan and Samar as well as in the Moluccas and Sulawesi. Locally known as *almaciga* or *almasigia*. This plant is possibly also in other parts of Southern Philippines. It is forest emergent and locally common, up to 1200 masl. Its conservation status is currently vulnerable [5]. Leaves are oval and acute, on vigorous shoots need to be narrower (**Figure 1**). It is known as a source of resin [6]. The resin was burned in the Cordillera Region of Northern Luzon that emit smoke that was inhaled to relieve bronchial asthma [7, 8]. The resins are sold in the Internet and known in its trading names of gold or black copal.

Agathis philippinensis Warb. is found in Babuyan Islands, Cebu, some parts of Luzon, and Mindanao, particularly in Bukidnon, Davao del Sur, and occurs mainly in montane forests to 2200 masl. *A. philippinensis* is a native species with a vulnerable conservation status [5]. Locally known as *almaciga*, it can be a source of resin, or Manila copal has been a valuable income source for indigenous peoples [9]. Various products can be made from wood, such as veneer and plywood, pulp and paper, and sources of construction materials [6].

Genus *Araucaria* Juss.

The genus is named after one of the provinces in Chile, Arauco [10].

Many species of *Araucaria* have been important sources of timber due to their massive size. *Araucaria* has two species not native to the Philippines, *Araucaria bidwillii* and *A. heterophylla*, which is more common.

Araucaria bidwillii Hook. is naturally distributed in Australia. Its large seeds are nutritious and was an important food source for aboriginal peoples and remain a popular delicacy.

Araucaria heterophylla (Salisb.) Franco. originated from Norfolk Island, an island territory of Australia cultivated, not naturalized, and not native in the Philippines. *A. heterophylla,* also called Norfolk Island pine, is used as a Christmas tree and whose leaves are used in Advent wreath.

Figure 1.
Leafy shoot of Agathis dammara (©MK Torres).

3.2 Family Cycadaceae

Genus *Cycas* L.

The genus name was derived from the Greek *koikas*, which was used by Theophratus to describe a type of palm. Later, it was transliterated to *kykas* and thence became *Cycas*. *Cycas* is the most primitive cycad genus.

All plant parts are poisonous, but people learned how to avoid this and its seeds are used as a food source. Such use is documented throughout the range of *Cycas* and its relatives in some parts of the world. In the natives of Africa, Australia and other countries, stem starch or *sago* has been used as a food source. Poisons derived from its parts have used to capture fish in Southeast Asia. Leaves are used as a substitute for palm fronds. Most species are popular ornamentals.

In Palawan, various *Cycas* spp. are found, like *Cycas aenigma* K.D.Hill & A. Lindstr., a native, endemic species known only in cultivation in Puerto Princesa. Its conservation status is critically endangered [5].

Another endemic species, *Cycas curranii* (J.Schust.) K.D.Hill. is confined to lowland ultramafic rocks, occurring in the understory of mixed closed forest, common on steep slopes, and occasionally on alluvial outwash from ultramafic hills in Palawan. Native and its current conservation status is critically endangered [5].

A critically endangered [5] known only from St. Paul's mountain massif, in crevices of vertical limestone cliffs with no soil, *Cycas saxatilis* K.D.Hill & A. Lindstr. is a significant outcrop to the Southeast, pendulous on limestone cliff face. In Culion Island,

Cycas wadei Merr. is found growing with seasonally dry *Imperata* grassland periodically subjected to fire. It is endangered [5].

Cycas nitida K.D.Hill & A. Lindstr. is an endangered species in the littoral forests of Rapu-rapu and Polillo Islands of Quezon Province and some parts of Luzon.

Endemic to Zambales, Luzon island is *Cycas zambalensis* Madulid & Agoo which grows in seasonally dry hilly grassland over chromite-rich ultramafics, scattered throughout but critically endangered [5].

Other endemic species, namely *Cycas riuminiana* Porte ex Regel. found in the understory of lowland rain forests, forested areas on ridges and mountains, in closed mixed evergreen forests usually on steep slopes, and disturbed areas, often on limestone, 0–1030 masl.

Cycas vespertilio A.Lindstr. & K.D.Hill grows in hill forests seasonally deciduous forests; both are vulnerable in conservation status and are distributed in some provinces in Luzon and Visayas islands.

In Mindanao, *Cycas flabellata* Agoo, Madulid & J.R. Callado is an endemic species of Bukidnon.

Cycas lacrimans A.Lindstr. & K.D.Hill., is found in forests over ultramafic rocks on serpentine soil in Davao Oriental. It is now endangered [5].

Another native, *Cycas mindanaensis* Agoo, Madulid & J.R. Callado, an endemic also found in Davao Oriental.

A cycad endemic to Mindanao, *Cycas sancti-lasallei* Agoo & Madulid found in Misamis Oriental, Cagayan de Oro, Cugman River Watershed found in disturbed lowland forests is critically endangered [5]. Interestingly, *C. sancti-lasallei* is named by its discoverers after the Saint John Baptist de La Salle.

Cycas edentata de Laub. is distributed in Luzon and Mindanao Islands, Philippines, but is also found in Andaman Islands, Borneo, Java, Lesser Sunda Islands, Malay Peninsula, Sulawesi, Sumatra, Thailand. This vulnerable species [5] thrives in coastal vegetation and lowland rainforests immediately behind it and rarely ventures far from the coast.

The most common but not naturalized, though cultivated in the country, is *Cycas revoluta* Thunb, used in traditional landscaping and popular during Palm Sunday every Lenten season.

3.3 Family Gnetaceae

Genus Gnetum L.

Gnetum is an insect-pollinated plant. A dioecious, both male and female strobili are aromatic that vary from with species. The strobili opens in the morning, while some species in the evening. The female strobilus produces a droplet of sticky sugary fluid that in time retracts into the strobilus, carrying any captured pollen to the nucellus. This extraordinary mechanism thrives best in high humidity conditions and has been seen as one of the reasons why *Gnetum* is restricted to rainforests.

Recent ethnopharmacological studies have found C-glycosyl-flavones, while a group of complex stilbenes and stilbene-substituted benzofurans are present.

Gnetum arboreum Foxw. is endemic to the Philippines and is distributed in Camarines Sur, Nueva Ecija, Quezon, and Zambales in Luzon.

The most common species is *Gnetum gnemon* L., which is found in Bataan, Batangas, Bulacan, Cagayan, Camarines Norte, Camarines Sur, Isabela, Kalinga, Laguna, Quezon, Rizal, Sorsogon in Luzon Island and Agusan del Norte, Davao, Davao Oriental, Davao del Sur, Lanao, Misamis Occidental, South Cotabato, Surigao, Surigao del Norte, Zamboanga, and Zamboanga Sibugay in Mindanao in the Philippines. Female flowers are shortly tipped as compared to axillary male flowers (**Figure 2**).

Figure 2.
Leafy branch of Gnetum gnemon showing female ♀ and male ♂ inflorescence (© MK Torres).

In other countries, its inner bark was used for fiber while some parts are edible, where it is cultivated as a fruit tree.

One variety, *G. gnemon* var. *gnemon,* is found in Fiji and Solomon Islands to Malesia, from Sumba and Sulawesi through the Philippines to New Guinea, the Malay Peninsula, and possibly elsewhere, often planted and frequently naturalized in secondary forests, even in the western part of Malesia.

The specific epithet *gnemon* is from the word *genemo*, the species' vernacular name in the Moluccas [11].

Gnetum gnemonoides Brongn. is a native species distributed in the Bismarck Archipelago, Borneo, Java Sea, Malay Peninsula, Moluccas, New Guinea, and Sulawesi. In the Philippines, it is found in Quezon, Zambales, and Agusan del Norte in Mindanao.

Gnetum latifolium Blume is found in many provinces in north of Luzon, some parts of Visayas, and areas in Mindanao. This native species has four recognized varieties: var. *latifolium*., forma *latifolium*, and var. *longipes* (Markgr.) T. H. Nguyên, forma *longipes* Markgr., var. *laxifrutescens* (Elmer) Markgr. and var. *minus* (Foxw.) Markgr. It was documented that the bark fiber of *G. latifolium* var. *minus* is used in making ropes and nets [6].

3.4 Family Pinaceae

Genus *Pinus* L.

The generic name *Pinus* was the Roman name for pine. *Pinus* is one of the oldest extant conifer genera and are economically important as timber, pulp, tar, and turpentine. For many years, it is a principal source of timber for many purposes, including firewood, construction, woodworking and others. Leaves in bundles of three needles (rarely two) differs *Pinus kesiya* (**Figure 2**) from *Pinus merkusii* with leaves in bundles of two needles (**Figure 3**).

Pinus kesiya Royle ex Gordon is found in Albay, Aurora, Benguet, Ifugao, Ilocos Norte, Mountain Province, Nueva Ecija, Nueva Vizcaya, Pangasinan, Zambales Mindoro and Negros Island. One variety, var. *langbianensis* (A. Chev.) Gaussen ex Bui is distributed in Albay, Benguet, Ifugao, Mountain Province, Nueva Ecija, Nueva Vizcaya, Pangasinan, Zambales that often occurs in pure stands, often on steep slopes, (300-)700–2700 masl. This native is forming a natural hybrid in Zambales. Female cones are dried and used as Christmas accent décor. Timber is used in various materials in construction. Oleoresin is an important component of insect-repellent products [6]. Pine needles were gathered on top of graves by locals of Sagada, Mountain Province remembering the dead during November 1. An annual tradition called *Panag-apoy*, which in *Kankana-ey* (South-Central Cordilleran language) means 'to light a fire'.

Pinus merkusii Jungh. & de Vriese can be seen in the provinces of Bulacan, Rizal, Zambales, and Mindoro. This vulnerable [5] and native pine has one recognized subspecies, ssp. *merkusii* is found in China, Indochina, Myanmar, Sumatra, and Thailand. They thrive in strongly seasonal areas from 100 to 2000 masl. In the Philippines, turpentine was once extracted from this species. However, the extraction rate was lesser than that of *P. kesiya* [12, 13]. It is a source of oleoresin. Wood is used in construction and pulp for paper manufacture [6]. The resin is generally higher in α- than β-pinene, which makes it contribute to production being used primarily for internal consumption for solvents rather than for export to the expanding aroma industry (**Figure 4**) [9].

Figure 3.
Pinus kesiya showing (1) sterile twig (2) mature female cone (3) bundle of needle leaves in 3's (© MK Torres).

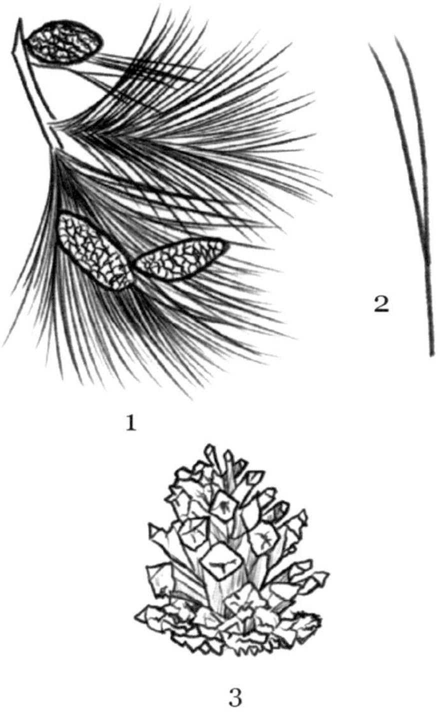

Figure 4.
Pinus merkusii showing (1) twig with young female cones (2) pair of needle leaves (3) mature female cone (© MK Torres).

3.5 Family Podocarpaceae

Genus *Dacrycarpus* (Endl.) de Laub.

The genus name came from the Greek words meaning 'tears' and 'fruit', describing the shape of its seed cones. *Dacrycarpus* spp. are dioecious shrubs or trees.

Both native species, *Dacrycarpus cumingii* (Parl.) de Laub. and *Dacrycarpus imbricatus* (Blume) de Laub. are found in some provinces in Luzon and Mindanao as well as in neighboring countries of the Philippines. *D. imbricatus* is another threatened species [5]. and has two varieties, namely, var. *imbricatus*. is common in montane cloud forests, 1300–2500 masl, and var. *robustus* de Laub. Wood is used to manufacture many products and can be a potential source of pulp for paper [6].

Genus *Dacrydium* Sol. ex G.Forst.

Dacrydium is derived from the Greek word which means 'tear', which refers to resinous exudations from the wood, [14, 15] but seemingly to the weeping habit of the trees [16].

Wood is yellowish to reddish in color, very resinous and known for its durability. It is used for building purposes, furniture making and woodworking [14].

This genus has four species found in certain provinces of Luzon and Mindanao, namely, *Dacrydium beccarii* Parl. in DC. *Dacrydium elatum* (Roxb.) Wall. ex Hook; one identified as other threatened species [5]. *Dacrydium pectinatum* de Laub., and *Dacrydium xanthandrum* Pilg. which is known from one collection from Mindanao also thrives in neighboring country islands of the Philippines.

Genus *Falcatifolium* de Laub.

Species are dioecious shrubs to large trees with thin smooth brownish bark with scattered lenticels, reddish and somewhat fibrous within. They are loosely and irregularly branched trees. No ethnobotanical information was documented.

Falcatifolium gruezoi de Laub. is a native species found in exposed sites along ridges or on borders of open areas, 1600–2200 masl. is distributed in Moluccas and Sulawesi and some parts of Luzon, Panay Island, and Mindanao.

Genus *Nageia* Gaertn.

The genus name is a Latinized form of ナギ (*nagi*). Some species are exploited for its high oil content of their seeds aside from its valuable wood.

One native species is *Nageia wallichiana* (C. Presl) Kuntze found in Borneo, China, India, Indochina, Java, Lesser Sunda Isls, Malay Peninsula, Moluccas, Myanmar, New Guinea, Philippines, Sulawesi, Sumatra and Thailand. In Luzon, they are distributed in the provinces of Apayao, Bataan, Benguet, Cagayan, Laguna, and Nueva Ecija, while in Mindanao in Misamis Occidental and Surigao del Sur. This other threatened species [5] is also found in Mindoro, Panay, Samar, and Sibuyan islands.

Genus *Podocarpus* L'Hér. ex Pers.

Podocarpus is derived from two Greek words *poús*, 'foot' and *karpós*, 'fruit' referring to its fleshy fruit stems. *Podocarpus* species are evergreen shrubs or trees with very limited distribution, often found at high elevation, and the majority of taxa have no recorded human use. However, human use is usually confined to relatively large individuals that thrives in proximity to populated areas.

Ethnopharmacological uses of leaf and bark extracts that treat a variety of ailments are reported for many species. Wood is strong, straight-grained and can use for construction, boat-building, furniture and household tools.

Endemic to the Philippines are *Podocarpus lophatus* de Laub., a vulnerable species [5] and sometimes shared in cloud forests, 2000–2100 masl.

Podocarpus macrocarpus de Laub. is distributed in Benguet, Ilocos Norte, Mountain Province, Quezon, Zambales, and Davao Oriental. Wood from *P. macrocarpus* can be used in airplane construction and manufacture of sounding boards, pencil slats and tennis racket handles [6].

Podocarpus palawanensis de Laub. & Silba found only in Palawan is critically endangered [5].

Podocarpus costalis C. Presl. is an endangered species [5] found in coastal bluffs near sea level to at least 300 m in Babuyan Islands, Batanes, Ilocos Norte, Isabela, and Polillo Islands.

Other native species include *Podocarpus glaucus* Foxw., *Podocarpus neriifolius* D. Don in Lamb., *Podocarpus pilgeri* Foxw., *Podocarpus polystachyus* R. Br. ex Endl. and *Podocarpus rumphii* Blume were declared vulnerable [5]. The wood of *P. rumphii* has economic potential for its local community [6].

Podocarpus ramosii R. R. Mill., found in Camarines Sur, Laguna, and Quezon, thrives in stunted mossy forests up to 2200 masl. is endangered [5].

Genus *Sundacarpus* (J.Buchholz & N.E.Gray) C.N.Page.

Sundacarpus amarus (Blume) C. N. Page is a native species in Australia, Borneo, Java, Lesser Sunda Islands, Moluccas, New Guinea, Sulawesi and Sumatra. It is distributed in Benguet, Ifugao, Ilocos Sur, Mountain Province in Luzon, and Davao del Sur in Mindanao. It is locally important for lumber [6].

Leaves are reported as bitter, to which also the Sundanese name '*pait*' refers, bittersweet '*dulcamara*', or sweet tasting. Thus, the specific epithet is from the Latin *amarus*, 'bitter', referring to its leaves.

3.6 Family Phyllocladaceae

Genus *Phyllocladus* Rich. ex Mirb.

Phyllocladus is differentiated from podocarps (formerly under one family) with a structure resembling an aril and possessing an epimatium, aside of having a different number of chromosomes and distinct pollination mechanism. It was documented that almost all species have been exploited for their timber although no conservation status is reported.

Phyllocladus hypophyllus Hook. f. is a native species that grows in the mossy forest on the higher mountains, 1200–2400 masl. in Borneo, Moluccas, New Guinea and, Sulawesi. In Luzon, it is distributed in Abra, Benguet, Ifugao, Ilocos Sur, Isabela, Mountain Province, Nueva Vizcaya, Mindoro, Palawan, and Sibuyan Island. In Mindanao, it thrives in Agusan del Norte, Bukidnon, Cotabato, Davao, Davao del Sur, Lanao, Misamis Occidental, and Zamboanga del Sur. Leaves spaced along branches, spiral and simple (**Figure 5**).

3.7 Family Taxaceae

Genus *Taxus* L.

This genus is named from the Greek *toxus*, reflective of *tóxon*, 'bow' and *toxikon*, 'poison'. As its extract was used as an arrow poison [17].

Commonly known as yew, *Taxus* spp. had many interesting uses in both traditional and modern culture, and its most noteworthy use is being a tree of profound spiritual significance. Nowadays, yew is known for being able to cure cancer [18]. Specifically, *paclitaxel*, a compound found in its vegetative parts, provides a very effective treatment for breast and ovarian cancer, and even reported in treating certain other cancers. Its discovery in the 1990s led quickly to its worldwide overexploitation to extract paclitaxel from its bark and foliage, making a number of species toward endangered status.

Figure 5.
Phyllodes of Phyllocladus hypophallus (© MK Torres).

Figure 6.
Twig of Taxus wallichiana (© MK Torres).

The only *Taxus* sp. found in the Philippines is *Taxus wallichiana* Zucc. It is located in moist subtropical forests, tropical highland ridges, and mossy forests in the canopy and is locally dominant; 1400–2300 masl in Benguet, Laguna, Mountain Province, Quezon, and Davao. Leaves are linear to linear-lanceolate (**Figure 6**). It is vulnerable (described as *Taxus sumatrana* (Miq.) de Laub. [5]).

4. Discussion

Unlike many other countries, Philippines is an archipelago with "tropical rainforest climate" rich with gymnosperm biodiversity. **Figure 7** shows the distribution

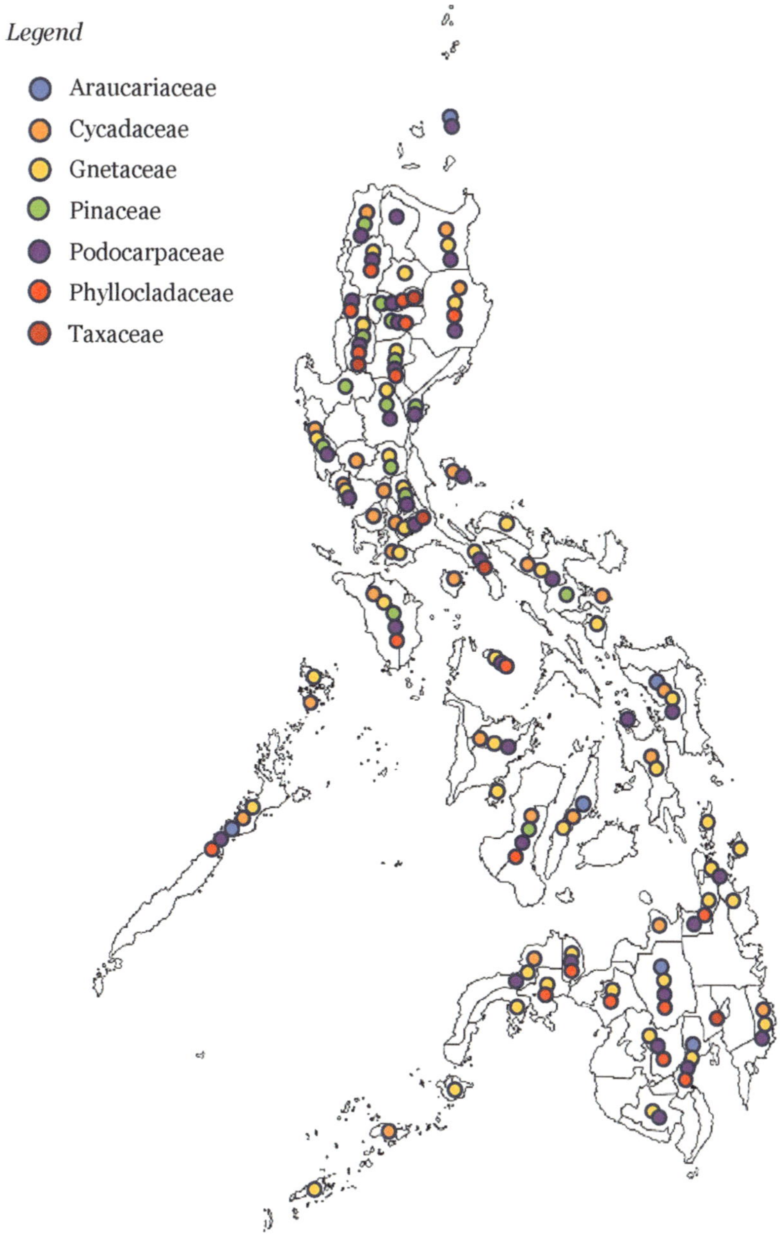

Figure 7.
Distribution map of gymnosperms in the Philippines.

of gymnosperms in the country. It is noticeable that some provinces showed the absence or few species from the different families. It was known that many species are rare and remote. *Agathis dammara* is considered the most widespread found in major islands of the Philippines. *Pinus kesiya* and *Pinus merkusii* have been observed inhabit forests areas of considerable extent. *P. kesiya* occurs primarily in the Cordillera Mountain range in Northern Luzon while *P. merkusii* is found mainly in Zambales, Luzon and Mindoro Island. Cycads occur across the country and thrive relatively at low elevations, relatively sparse in comparison to Indochina and northern Australia, both of which show extensive radiations [19]. *Gnetum* species are found in tropical forests. The Philippines is considered as one of the centers of diversity for the Podocarpaceae along with nearby geographical regions [20]. Geographical distribution of taxonomic groups remains poorly documented. On a global scale, only broad conclusions about the patterns of gymnosperm diversity and its vulnerability can be formulated [21].

Among the presented species, only few conifers have economic use. In the presented data, *A. dammara, Agathis philippinensis, Gnetum gnemon, P. kesiya, P. merkusii* and *Dacrycarpus imbricatus* exhibit the outstanding economic potentialities. The rest of the gymnosperms have no information available.

5. Conclusion

The ethnobotany of conifers in the Philippines showed few or little information. The Philippines is evidenced with diverse gymnosperms. However, it is necessary to harness the economic potential of conifers widely distributed in the country. Research in the field of ethnobotany particularly in this group of plants still needs to be filled. Immediate analysis of its distribution and biodiversity conservation of these group of plants should be established.

Acknowledgements

All line drawings are done by Ms. MK Torres (2023).

Author details

Richard Clemente
Bulacan State University, City of Malolos, Bulacan, Philippines

*Address all correspondence to: richard.clemente@bulsu.edu.ph

References

[1] The Gymnosperm Database: Home Page. Available from: https://www.conifers.org/index.php

[2] Barcelona JF, Nickrent DL, LaFrankie JV, Callado JR, Pelser PB. Co's digital flora of the Philippines: Plant identification and conservation through cybertaxonomy. Philippine Journal of Science. 2013;**142**:57-67

[3] Boland DJ, Brooker MI, Chippendale GM, Hall N, Hyland BP, Johnston RD, et al., editors. Forest Trees of Australia. Melbourne, Australia: CSIRO Publishing; 2006

[4] Whitmore TC. A First Look at Agathis. Tropical Forestry Papers No. 11. Oxford: Commonwealth Forestry Institute, University of Oxford; 1977

[5] DENR Order No. 11 of 2017 (Updated National List of Threatened Plants and their Categories). Department of Environment and Natural Resources. 2017

[6] Zamora PM, Co L. Economic ferns; Endemic ferns: Gymnosperms. In: Guide to Philippine Flora and Fauna. Quezon City, Philippines: Natural Resources Management Center; 1986

[7] Pennacchio M, Jefferson L, Havens K. Uses and Abuses of Plant-Derived Smoke: Its Ethnobotany as Hallucinogen, Perfume, Incense, and Medicine. New York: Oxford University Press; 2010

[8] Co LL. Common Medicinal Plants of the Cordillera Region (Northern Luzon, Philippines): A trainer's Manual for Community-Based Health Programs. Baguio City: Community Health Education, Services in the Cordillera Region; 1989

[9] Langenheim JH. Plant Resins: Chemistry, Evolution, Ecology, and Ethnobotany. Portland, Oregon, USA: Timber Press; 2003

[10] Vidaković M. Conifers: Morphology and Variation. Zagreb, Croatia: Grafičko Zavod Hrvatske; 1991

[11] Markgraf F. Gnetaceae. Vol. 4(3. Djakarta: Noordhof-Kolf; Flora Malesiana Series 1; 1951. pp. 336-347

[12] Mustaqim WA. *Pinus merkusii* Jungh. & de Vriese Pinaceae. Ethnobotany of the Mountain Regions of Southeast Asia. Switzerland: Springer Nature; 2021. pp. 881-888

[13] West AP, Brown WH. Phillippine Resins, Gums, Seed Oils, and Essential Oils. Manila, Philippines: Bureau of Printing; 1920

[14] Dallimore W, Jackson AB, Harrison SG. A Handbook of Coniferae and Ginkgoaceae. 4th ed. (rev.) ed. New York: The New York Botanical Garden; 1966

[15] Farjon A. A Handbook of the World's Conifers. Vol. 2. Leiden, The Netherlands: Brill; 2010

[16] Kirk T. The forest flora of New Zealand. G. Didsbury, Government Printer; 1889

[17] Hartzell H. The Yew Tree: A Thousand Whispers: Biography of a Species. Hulogosi Communications, Incorporated; USA: The University of Michigan; 1991

[18] Bryan J, Twelves C. How bark from the Pacific yew tree improved the treatment of breast cancer.

The Pharmaceutical Journal. 2011;**287**(7672):369

[19] Lindstrom AJ, Hill KD, Stanberg LC. The genus *Cycas* (Cycadaceae) in the Philippines. Telopea. 2008;**12**(1):119-145

[20] Khan R, Hill RS, Liu J, Biffin E. Diversity, distribution, systematics and conservation status of Podocarpaceae. Plants. 2023;**12**(5):1171

[21] Fragnière Y, Bétrisey S, Cardinaux L, Stoffel M, Kozlowski G. Fighting their last stand? A global analysis of the distribution and conservation status of gymnosperms. Journal of Biogeography. 2015;**42**(5):809-820